www.bitethepaintbrush.com

The Mendelevian Oracle

ISBN 979-8-9865246-1-0

A Guide to the Science and Mysticism thereof

By Robin McFadden

Foreword

Our journey begins, as they so often do, with a shower curtain.

In 2015 my family bought a periodic table-printed shower curtain, and for a while I was satisfied. It was nerdy and quirky and cute. It didn't matter that I did not know the individual elements beyond the basics like "Hey, oxygen! We breathe that!" But eventually I began to wish I knew a little more, and I longed to have the kind of brain that doesn't delete dry facts. And then one day I thought: If they were *characters*, it would be easy to remember everything about them…

And the image of Hydrogen, pregnant and armed, dropped cold into my head.

From that moment it became my mission to design fantasy creatures based on every single one of the 118 elements of the periodic table. Not only that; to assign each of them a chemically appropriate intangible quality so that, once printed on cards, the whole crew could be used to learn chemistry OR to read fortunes. I learned a ton. I was deeply immersed. I barely noticed lockdown. And whenever I began to think I was an idiot for trying to combine chemistry and mysticism, I went and reread the quote I had found by sheer good luck near the beginning of this whole saga—a quote that has given me hope that all my floopy weirdness will matter to anyone besides me:

It is probably quite true generally that in the history of human thinking the most fruitful developments frequently take place at those points where two different lines of thought meet. These lines might have their roots in quite different parts of human culture, in different times or different cultural environments or different religious traditions: hence if they actually meet, that is, if they are at least so much related to each other that a real interaction can take place, then one may hope that new and interesting developments may follow.

-Werner Heisenberg

This gave me the strength to silence the intrusive voices in my head. I painted for eighteen months, and wrote for three years. As I said, I learned a lot; and whenever I got stuck, things and people conspired to pull me forward again. Big thanks go to my family, who gave me space, modeled for several poses and never complained that the wall in the hallway was covered in unfinished paintings for a year and a half. My dad cloud-seeded the lefthand science essays for me when I was stuck. And I owe an enormous debt of gratitude to Theodore Gray, whose book *The Elements* was my map and compass—and he also provided the spectrum emission lines on the cards.

Speaking of the cards, there is a wealth of information on them, as demonstrated in this handy example:

Atomic Number
Number of protons, also number of electrons.
Defines its place in the table.

Emission Lines
When an element is heated, it emits light in very specific wavelengths.
Like a fingerprint.

Atomic Symbol
Shorthand form, used in chemical formulae. Every painting has its symbol in there somewhere.

Crystal Structure
A pure sample of an element tends to arrange itself in a specific pattern. Not a hard-and-fast rule, but useful to know.

Bohr Model
Niels Bohr came up with the idea of electrons existing in layers of orbital shells, like planets in a solar system. This is still a useful image to use, even though it is an oversimplification.

Element Name and Oracular Meaning
These are fairly self-explanatory. The oracular meanings were chosen to reflect each element's traits, uses, or history.

Atomic Weight
As the number of neutrons varies from atom to atom (these variations within each element are known as isotopes), a pure sample of an element has a mass that is more than the sum of the protons, but rarely a whole number; as the weight is achieved by averaging the common isotopes of that atom.

There are also certain "codes" embedded in the paintings that may help you remember their traits. For instance:

(skip the rest of this page if you wish to discover them for yourself)

The background colors show which period an element is in. Top to bottom: black, purple, blue, turquoise, green, brown, grey.

The gases all have wings; the noble gases are dragons. Those which are liquid at normal temperature and pressure are merfolk. The alkali metals are trolls and the alkali earth metals are goblins; everything radioactive is glowing, everything synthetic is cyberpunk. The rare earths are all amphibians: the lanthanides are frogs and the actinides are toads. Those which have only existed for nanoseconds—whose chemical traits cannot be observed—are wearing masks. (Thus oganesson, the final card, is a glowing masked cyberpunk dragon.)

Since there are a lot of similarities within the columns (called groups), I tried to give them physical similarities as well, but not in a consistent way. After all, the elements are all ultimately individuals, describable by location but not defined by it. So the boron group is all girls with long white hair, and the oxygen group (or chalcogens) are all satyrs…but the nitrogen group mostly just share a facial expression, and the zinc group are just sort of…not anyone else.

And while we're talking D block elements, it was awfully tempting to make all of the transition metals into dwarves; but that would have gotten monotonous, so instead I made the groups easy to separate. They are, from left to right: amphibians, reptiles, marsupials, mammals (all these are anthropomorphic), hobbits, dwarves, gnomes, leprechauns, elfin royalty and...sort of elves (that's the zinc group again). And finally, for those paying attention, you will notice that the "platinum group" (not a vertical group) all have hexagon patterns on their clothing.

I am sure you will spot other patterns. I hope you find all this useful and inspiring. I hope complex molecules start to look like adventuring parties to you, and that your further education about the building blocks of the physical universe is made easier by these creatures living in your head. I also hope that when you use the oracle, it brings you wisdom and clarity and joy.

I hope that new and interesting developments may follow.

-Robin McFadden

December 2022

This book is a guide to both the science and the oracle. Each page spread has pure factual information on the left, and a more magical interpretation on the right. Just find the number of the element you need in the upper right corner, and go.

Learn things. Blessed be.

Hydrogen

Element Number:	1
Weight:	1.00784u
Normal State in Nature:	gas
Melting Point:	13.99 Kelvin
Boiling Point:	20.271 Kelvin

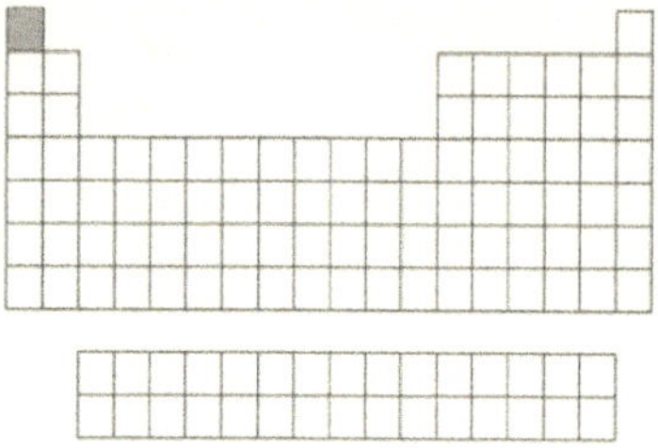

Henry Cavendish is credited with hydrogen's discovery in 1766. In 1783 Lavoisier named it for the Greek *hydro* "water" and *gen* "maker" since water is a product of burning hydrogen.

Hydrogen-filled airships were a promising technology until 1937, when the explosion of the Hindenburg rendered them unpopular, to put it mildly. Hydrogen is still used to float weather balloons, though. And hydrogen was key to atomic theory in the early twentieth century, as it still is. It's the only element that Schrodinger's equation can be shown to work on.

It is possible to spend your entire lifetime studying hydrogen, and a lot of scientists have done just that.

One atom is instantly explosive. One molecule, consisting of two atoms, is stable as long as you leave it alone. It is the most abundant element in the universe, by a long shot. For all that hydrogen is the simplest element that exists, or even *can* exist, it is remarkable in what it can do.

Power stations use it as a coolant for their generators. Cryonics, petrochemical processing, and ammonia production all require it, and so does your own body. But if that doesn't sound impressive enough, remember that the Sun is a ball of burning hydrogen. Neither you nor I nor anything else would exist without it. Perhaps they should have called it "life-maker" instead.

1

Source

"I travel a great deal, but I never wander."
"Then I assume that you move about with direction. What is your usual destination?"
"The source. I am always voyaging back to the source."

-Tom Robbins, *Another Roadside Attraction*

Everything that exists, started out as hydrogen. When the universe spat itself out in the Big Bang, 13.7 billion years ago, the first atoms that formed were hydrogen. Eventually this collected into balls, which ignited, and stars began the massive project of using fusion to turn hydrogen into…everything else. Despite the many fusions and transformations it has undergone, much hydrogen has yet remained unchanged since the beginning. It is still the most abundant element of all.

Its position as the origin of all matter is awe-inspiring, but there is another aspect to this element, particularly where it relates to humankind. This lighter-than-air gas is highly explosive. There are a number of incidents in history when we failed to treat it with the proper respect, and paid a high price.

There are 118 elements that we have so far discovered or synthesized. Every one of them, before they were cooked into being in the furnace of stars, or by human ingenuity, was simply hydrogen. One proton, one electron. Simple, but in its simplicity lies the seed of the universe.

This card invokes the infinite possibilities of the Source. Everything you need exists here. Everything you are, came from here. It is in the air you breathe, in your blood, and in your dreams. If you seek counsel, if you feel that you are wandering without direction, center yourself and look within. You contain the universe. Be very respectful to this power you hold; you have the capacity to explode and burn, as well as the ability to create and sustain. They come from the same Source.

Helium

Element Number:	2
Weight:	4.002602u
Normal State in Nature:	gas
Melting Point:	1° Kelvin
Boiling Point:	4.2° Kelvin

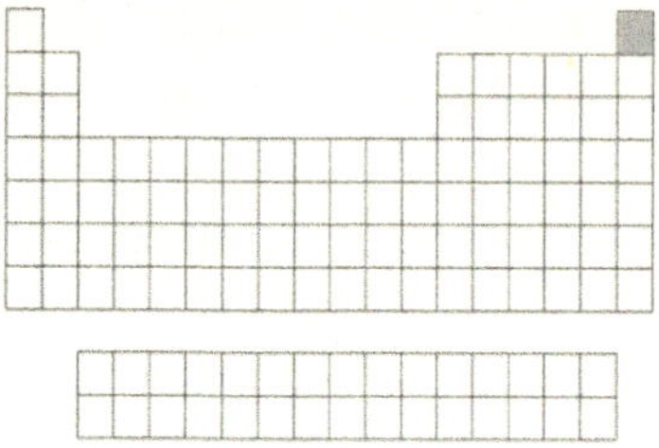

Helium's existence was first suspected when a solar eclipse in 1868 revealed a yellow spectral signature line to several keen observers. Lockyer named the new element after the Sun, but it was not formally discovered until 27 years later, by the chemists Ramsay, Cleve and Langlet.

Helium is the second most prevalent element in the universe, but rare on earth. Most of it was created one second after the Big Bang, and was produced by the decay of hydrogen into a stable atom, and a plasma. The decay of radioactive elements can also create it, deep in the Earth's crust. While it is naturally a gas, it can be compressed under massive pressure into a liquid, which is useful for transporting and storing it; and also into a solid, which isn't useful for much of anything.

You can't see it, smell it, or taste it. It won't burn, it won't explode, and it won't poison you, although it could suffocate you. Because it's so safe, it's a highly useful element.

Its biggest use? Making things cold. Very cold. Cryogenics and superconductors kind of cold. It's also used in welding. The Goodyear blimp is held aloft by helium. It is used as a tracer gas to detect leaks in vacuum systems, and as a filler gas in certain telescopes because of its low index of refraction. And, of course, there are the party balloons.

America currently has the biggest reserves of it, as it tends to mingle with natural gas. We can be free and easy with it if we so choose, but when it's gone, it's gone.

Adventure

"Of course I shall fly—does not Tara of Helium always do that which pleases her?"

-Edgar Rice Burroughs, *The Chessmen of Mars*

Helium is lighter than air. We can pull it out of natural gas and use it for various things, but it will always manage to escape and run away into the wild yonder. There is an enormous quantity of it in the universe, but it is invisible, odorless, tasteless, and very difficult to pin down. Like Adventure itself, it is a substance without substance, a thing that you find by following signs, and never expect to behave normally. When humanity does find Helium, we use it for serious science and industry: In superconductors, in cryogenics and in arc welding…and then again, we use it to fill party balloons, and give ourselves hilarious high cartoon voices. The largest and loftiest destinations are intimately tied to simple, childlike delight.

A warning, however: Inhaling helium is a joy every small beast should experience, as we know; but if we do it too much, we will suffocate. In the same way, a diet of pure adventure cannot be sustained. A life made this way will drift off into nothing. Tempered with other, more prosaic experiences, we gain balance. We can go on exploits, or have a comical chipmunk voice, as often as we please.

\When Helium appears, treat it as an invitation into the unknown. That yellow spectral line is a brick road through Oz, or a golden thread in a goblin cave, or a canary that sings in the forest, leading you onward. Venture fearlessly into the wilderness, following signs from the universe. Meet strange people and do entirely new things. Change your perspective and face fresh challenges. As long as you keep your bearings, and remember where Home is, you will come to no harm.

Fly on the gossamer wings of Adventure.

Lithium

Element Number:	3
Weight:	6.941u
Normal State in Nature:	solid
Melting Point:	454° Kelvin
Boiling Point:	1615° Kelvin

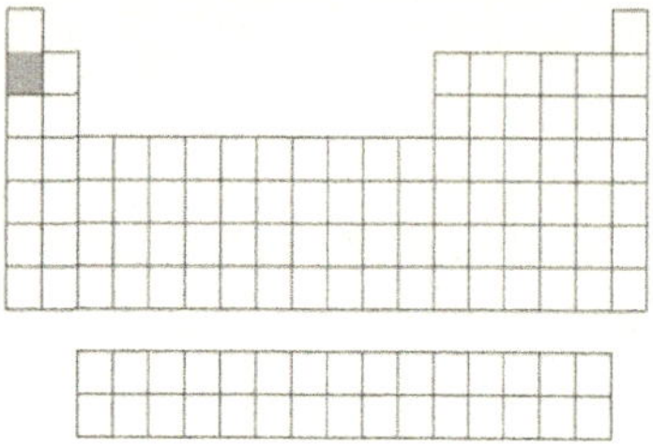

It was not terribly common to detect a new element within a mineral, so when Arfwedson discovered a previously unknown alkali metal in petalite ore in 1817, his mentor Berzelius named it "lithion" after the Greek word for "stone." It was successfully isolated from its salts in 1821.

It was produced during the Big Bang and is still produced during the creation of new stars. Most lithium on earth is in ocean water. It does not occur in nature as a pure element, only as a salt. It is extracted from those salts in a variety of ways. Once pure, it has the density of soft pine, and is the softest of the solid elements. Its popularity has fluctuated over the past hundred years, influenced by difficulty of extraction, advances in medical science, Cold War demands, and battery technology.

It is one of the most useful elements, for everything from batteries to lubricant that doesn't melt (as petroleum lubricant does) in the presence of heat. It is used to treat anxiety and bipolar disorder, and there is some evidence that it increases chemical messenger activity in the brain. It can store and release energy. It lowers the melting point of glass. Lithium is technically a metal, but is so soft you can cut it with a butter knife. It can float on top of water. Don't try that without first protecting yourself, though. It is extremely unstable as a pure metal, and will explode and burn with red flames in contact with moisture. An overdose can damage your kidneys and semipermanently alter your level of consciousness.

3

Serenity

"Light my candles in a daze, 'cuz I've found God."

-Nirvana, *Lithium*

A metal light enough to float on water and soft enough to cut with scissors and smooth enough to grease up engines is inspiring indeed. Add to that lithium's capacity to quietly store energy until it is needed, its ability to purify air, and most of all its soothing effect on the human brain, and we have the guru of the periodic table, a chemical bodhisattva.

As with all elements, lithium is not an unalloyed font of good things. Though it is light enough to float on water, this is not something you want to witness unprepared, because it will start to burn and release toxic gas. It is also corrosive to skin, and must be handled accordingly. A lithium battery can, if made incorrectly, overheat and explode. All of these things also apply to our own sense of inner peace: When we force it, or demand it, or treat it cavalierly, it corrodes. Or explodes. And so do we.

But when we act with intelligence and patience, Serenity is ours.

Lithium is telling you that the time has come to cultivate your inner garden. Meditate. Bask in sunbeams. Take a nap. Take a break from social media. Hydrate. Read a book you already love. When you're tempted to put yourself last--push yourself harder—envision lithium getting dumped into water and burning away into poisonous fumes. That's you. And then envision lithium in its many uses: Soothing people, smoothing situations, holding energy and releasing it appropriately. That's you too.

As long as you allow yourself Serenity first.

Beryllium

Element Number:	4
Weight:	9.012182u
Normal State in Nature:	solid
Melting Point:	1560° Kelvin
Boiling Point:	3243° Kelvin

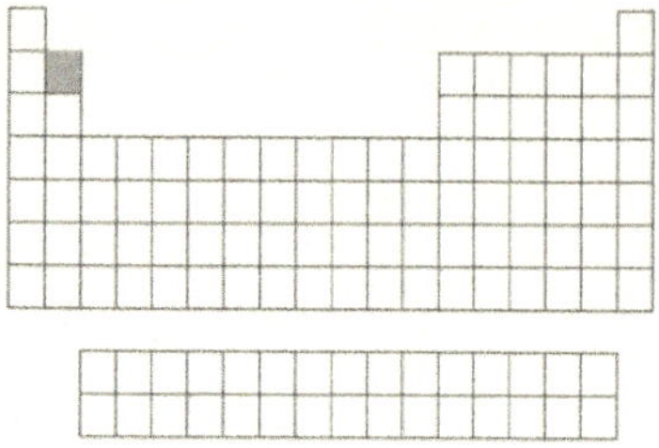

Beryllium ores and stones have been prized for jewelry and sculpture for some two thousand years, but only in the nineteenth century was it discovered as an element, simultaneously by Wöhler and Bussy. Previous to that, it was commonly held that the similarities between beryl and emerald were due to the presence of aluminum silicates. Wöhler was the first to call the new element beryllium, after the stone.

It wasn't until 1957 that pure beryllium was manufactured. Before that, its minerals were used to harden copper alloys. Once isolated, it becomes a very lightweight, rigid and thermally conductive metal. If it was not so rare and so toxic, it would be widely used. Even still, it is indispensable for aerospace, military, medical and optical applications.

In alloys, it hardens armor and artillery, as well as more frivolous things such as expensive golf clubs. It does not spark, so it is used in tools on oil rigs and in other high-fire-risk industries. It is totally transparent to X-rays, so about a third of all the beryllium produced is utilized for X-ray equipment.

It can kill you, which is the second worst thing it can do to you. It only dissolves in very acidic fluids, which means once it gets inside you, it doesn't come out. There's no treatment for berylliosis, which you get from breathing the dust when you're working with it.

Oh, and the worst thing berylliosis can do? Not kill you.

4

Sophistication

"The most sophisticated people I know - inside they are all children."

-Jim Henson

Strong and lightweight, invisible to X-rays, heat resistant, corrosion resistant and shiny; beryllium can be found in rockets, emeralds, missile parts…everywhere you'd expect a secret agent to go. And when it wants to kick back, beryllium can be found lending its strength to the copper alloy that is used in very expensive golf clubs. Wherever toxicity and high cost is allowable in exchange for incredible finesse, there be beryllium. Smiling that smile. Being deceptively tough.

Do you need non-sparking tools on your oil rig? Bring in the beryllium alloy. Nonferrous precision instruments to disarm a bomb? Call it in again. A thin yet strong window for your X-ray tube? Yes indeed. A super suave golf club to help you with your swing? Hello, it says with a smile. A sphere to power your fictional spaceship? Why not! Is its attitude.

Being sophisticated does not mean it doesn't get to have fun.

It is time to reveal your debonair side. You are James Bond: Adjust your cuffs and prepare to charm the world. You are Spaceman Spiff: Ride through the cosmos on the strength of your imagination. You are Captain John Carter: kicking ass while befriending Martians. Whatever guise you adopt, keep in mind the fact that, just as beryllium dust is quite poisonous, so can your cockiness get you into trouble.

So watch your ego while you drink that dry martini.

Boron

Element Number:	5
Weight:	10.811u
Normal State in Nature:	solid
Melting Point:	2365° Kelvin
Boiling Point:	4275° Kelvin

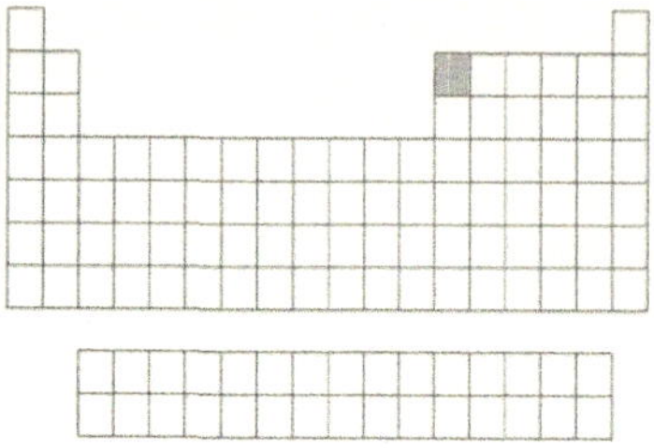

Though used in glazes and metalwork for hundreds of years, boron was first isolated by Davies in 1808, and not recognized as an element until Berzelius declared it so in 1824.

Boron is made by supernovas. That is where all of it comes from. As an element, it is a shiny black or dark brown metalloid. The only pure boron on Earth is what crashes on the planet.

It is a metal halide, and is used in almost every industry, from electronics to medicine, glassmaking to insecticides, fertilizer to homebuilding to fireworks. It is mined from the earth as a salt or a mineral compound, and most of it comes from California, which has the world's biggest active deposit, just outside the city aptly named Boron.

Cooking with Pyrex? Then you are taking advantage of the low coefficient of thermal expansion provided by borosilicate glass. Insulating the house with fiberglass? The structural integrity of those pink battens comes from boron. A glaze on your Ming Dynasty china? Boron helped it harden at a lower temperature.

It is very difficult to separate boron from its compounds, because the bonds it forms are so strong. It also captures neutrons quite effectively.

If, today, you do a load of laundry, or exterminate ants, or prevent a nuclear reactor meltdown, you will utilize boron. And it's really hard to poison yourself with it, although, if you drink a boron cocktail, you may feel like you're dying for a while.

5

Support

Old Jerome was a board-walk. Everybody knows that the world is supported by the shoulders of Atlas; and that Atlas stands on a rail-fence; And that the rail-fence is built on a turtle's back. Now, the turtle has to stand on something; and that is a board-walk made of men like old Jerome.

-O. Henry, *Schools and Schools*

In its pure form, which is never found in nature and difficult to isolate, boron is too brittle to have any uses. But in compounds and combinations, it is indispensable. Fiberglass, ceramics, polymers…Boron is never the star of the show, but if it were missing, the whole thing would quite literally fall apart.

Boron compounds strengthen plant cell walls: support for living beings. Boron oxide adds strength to fiberglass: support as insulation, protection, warmth. Borosilicate glass is resistant to thermal shock and gives us Pyrex: support for cooking your food. Boron carbide is in bulletproof vests, as well as nuclear power plants because it can absorb radiation without becoming radioactive: support of safety in extremes. Boric acid is an insecticide: support of the defense of the home. And of course, borax is for cleaning and silly putty. Support for hygiene and childhood.

Semiconductors use boron, neodymium magnets use boron…the world would fall apart without boron, but who knew that? Not many. Now you do.

When Boron arrives, take a moment to think of all the ways you improve the world without needing to be thanked. Sometimes we crave appreciation, but this card is telling you that today, achievement is its own reward. Let your desire for acknowledgment take a nap. Just get things done and be satisfied in your work, which is worthy and crucial and you don't need anyone to tell you that. There is peace to be found in the role of Support.

Carbon

Element Number:	6
Weight:	12.011u
Normal State in Nature:	solid
Melting Point:	3825° Kelvin
Boiling Point:	5100° Kelvin

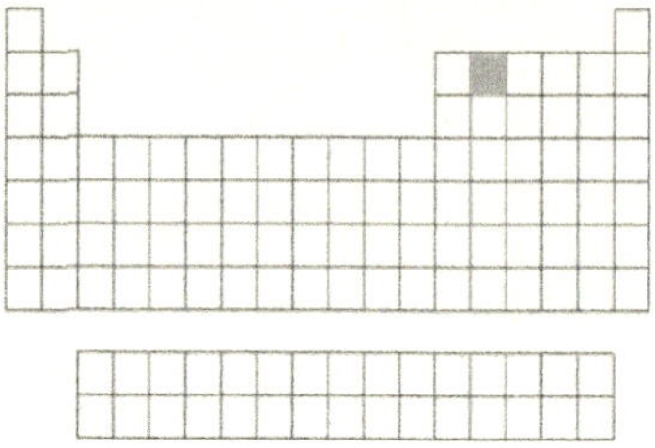

Known in various forms since antiquity, carbon is named for the charcoal humanity has been using since we discovered fire. Ötzi the Iceman has tattoos made with carbon that remain visible after more than 5000 years. Next to hydrogen, helium and oxygen, carbon is the most abundant element in the universe. It is second only to oxygen as the biggest component of our bodies. It has a vast array of uses, from the graphite in our pencils to the carbon fiber in our cars.

And it keeps being relevant. 18th century scientists found it in steel, graphite and diamonds, and defined it as an element in 1789. But carbon nanostructures such as buckyballs were invented as recently as 1985, and their uses have even now not been fully explored.

We heat with it and cool with it. We write with it, wear it, and we've found nearly ten million things it's part of. Carbon is the base for the world's most dangerous poisons: cyanide, sewer gas, mustard gas. But it is also the foundation of all DNA, and therefore all life. Diamonds, which are pure carbon, are the hardest known crystals. Diamonds will also burn up in a house fire. Ötzi's tattoos may be forever, but diamonds are not.

Carbon dioxide is what we, other animals, and a great deal of our modern infrastructure, exhales. The pre-industrial concentration of carbon in the Earth's atmosphere was 280 ppm. Now it is 390 ppm and climbing. Humanity would be wise to check this trend.

6

Life

"Maude, do you pray?"
"Pray? No. I communicate."
"With God?"
"With *Life*."

-Harold and Maude

Carbon's ability to form long-chain complex molecules makes it the perfect building material for DNA. Everything that contains DNA is dependent on carbon. We humans are carbon-based life forms, as scifi shows have been telling us for decades, but so is every living thing we know of, from the tardigrade to the giant redwood. Life itself is carbon-based.

And yes, pure carbon atoms can be stacked in a myriad of ways. Diamonds are carbon, and so is charcoal; graphite for your pencils, and "buckyballs," the beautiful arrangement of sixty carbon atoms that resembles a geodesic dome, and was therefore named in loving tribute to Buckminster Fuller, who invented them. And all of this is interesting; but our very ability to perceive things as interesting is a product of biology, which is ultimately a function of carbon. Isn't that the most interesting thing of all? And here's another interesting thought: Oxygen is necessary to humans, and carbon is indispensable; but when combined, they become gases that are harmful to us. Beware good things in bad combinations.

Drawing Carbon is a blessing. You embody the very essence of Life! Revel in it! Take a walk and really look around you. Acknowledge that every blade of grass, every animal and every tree is your sibling. We are all one, spinning on this good Earth together, made of the same stardust. You might find yourself singing Pagan songs and twirling in patches of sunlight. This is good. The more awake we are to the beauty of our world, the less likely we are to allow it to be destroyed by misplaced carbon. We belong alive.

Nitrogen

Element Number:	7
Weight:	14.0067u
Normal State in Nature:	gas
Melting Point:	63.15° Kelvin
Boiling Point:	77.344° Kelvin

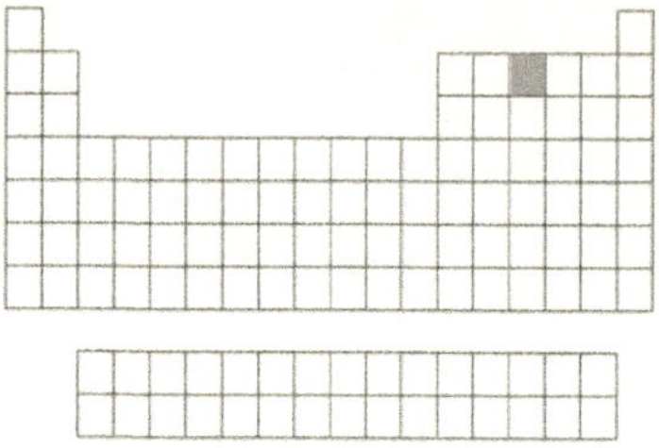

Ancient cultures used many nitrogen compounds. In the Middle Ages, chemicals containing it were the province of the alchemists. Science has always loved the way its acids dissolve things, but it was not until the 18th century that it was recognized as a gas; a gas which kills animals and extinguishes flames. It was called noxious (or burnt or mephitic) air; it is called *azote* in many languages and *stickstoff* in German, and finally, English-speaking scholars stole the Greek via the French and began calling it nitrogen in 1794.

Every single one of these names means some variation of "to choke."

Daniel Rutherford is given credit for the pure element's discovery, as of the three scientists who independently isolated it, he was the first to publish results.

Industrial processes that produce nitrogen-rich fertilizer completely changed the world in the early 1900s. Agriculture boomed, and so did munitions. Soil was thrown out of balance, as the nitrogen cycle got hijacked; but while it is true that its overuse can endanger public health, it is also true that half the world's agriculture depends on synthetically produced nitrogen. Its overuse can kill, but not using it is unthinkable.

Biologically, nitrogen is vital to life on Earth. Every nucleic acid, protein, and amino acid (to name a few) requires it. It carries energy to cells. Liquid nitrogen is incredibly cold and is commonly used in cryopreservation. Nitrogen can crowd the oxygen out of a room or a water system, sometimes lethally. It should never be taken lightly.

7

Price

"Come now," purred the Goblin girl. "Why do you malign my Goblin's wares? My silk is fine, come sip my wine, and are not my prices fair?"

-Catherynne M. Valente,
The Girl Who Fell Beneath Fairyland and Led the Revels There

Most of our atmosphere is inert dinitrogen, which is useless to humans. But when a man in the early 20th century discovered how to convert that into ammonia, which is a vital plant nutrient, the agricultural world went gaga and hasn't recovered yet. Farmers pour nitrogen fertilizer onto their fields, hoping for bigger crops. And they get them…at a price. Runoff from overfertilized fields gets into rivers and lakes, causing algae population explosions and suffocating everything else. Nitrogen in town water supplies leads to "blue baby warnings," a municipal response to the fact that, temporarily at least, giving your child tap water to drink will cause them to asphyxiate. Nitrogen fertilizer oversimplifies soil, leading to all kinds of problems. Such terrible waste and pollution…And yet, in certain forms, nitrogen is utterly necessary to humans, even at the level of our DNA. It's complicated.

Nitrogen has a natural cycle, from the air to the soil, from the soil back to air, nourishing us in some forms and killing us in others. Propellants, explosives, Kevlar and superglue all contain it, and the makers of these all know the price they would pay for carelessness, whether in innocence or ignorance. And the human body cannot function without nitrogen; it is vital to us, but it comes not free.

When there is nitrogen in your spread, educate yourself about where you may be (metaphorically) spending more than you can afford. The price of folly, perhaps? The price of ignorance? The price of selfishness—or is it the price of empathy, the price of overwork, the price of self-sacrifice? Check your ledger and make sure that you are paying the right price.

Oxygen

Element Number:	8
Weight:	15.999u
Normal State in Nature:	gas
Melting Point:	54.8° Kelvin
Boiling Point:	90.188° Kelvin

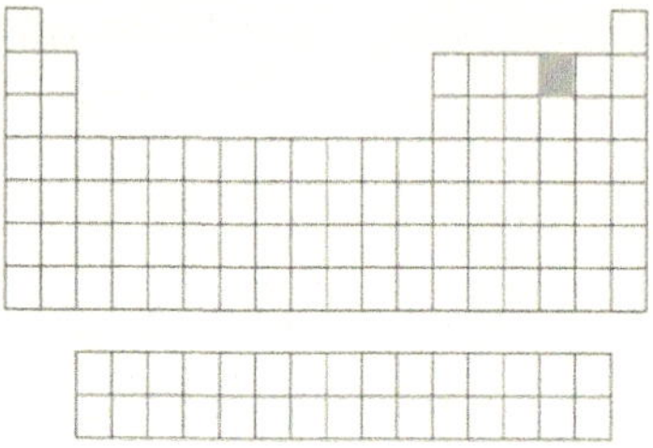

Oxygen is far too reactive to remain a free element for long. Fortunately for us, and all animals that breathe it, oxygen is continually replenished in the atmosphere by the photosynthesis of plants. It took a long time and many debunked theories before oxygen was observed as the key to combustion, and even longer before it was recognized as a chemical element; the first of the chalcogens.

Priestley and Scheele independently conducted important experiments with it in the 1770s, finally debunking the phlogiston theory which had been popularly accepted up to that point. Lavoisier in 1777 named it *oxygen* from the Greek for "mother of acid," mistakenly believing it to be a component of all chemical acids; and though he was wrong, the name is still impressively metal. For a gas.

Oxygen is the most reactive element in the entire table. It forms bonds with nearly everything. Because of its reactivity, and the fact that it is the most abundant element on Earth (though it's only number three in the universe at large), you will rarely find a molecule that doesn't contain it. Organic compounds require it. Not to mention water, without which there could be no life at all. It is also valuable as a rocket propellant, in the vacuum of space where the fuel would otherwise fail.

This is not to say that oxygen is purely benevolent. It makes things burn. If they're burning already, it makes them burn harder. It corrodes metals, rots food, and can even be toxic when breathed under pressure. Still, all things considered, probably worth it.

8

Sociability

"He'd be at home in Buckingham Palace or at the bottom of a coal mine," he said. "And yet it's not impudence, either. He's just fine, is that lad."

-Frances Hodgson Burnett, *The Secret Garden*

Oh, Oxygen! Would that we were all as good at playing with others as you!

Oxygen is our most social element. It forms bonds readily with virtually every other atom, not to mention a massive list of compounds. Good thing, too; oxygen is the most abundant element on Earth, so if it were an introvert, we would not be here to discuss it.

Oxygen is SO socially inclined, even when it's alone it forms bonds with itself. Dioxygen (O2) makes up 20% of our atmosphere. Ozone (O3) in our upper atmosphere protects us from UV rays. Bonded with hydrogen, it IS our oceans. It makes fire happen, not by being fuel, but by allowing fuel to burn.

Looking inward, we see oxygen delivering nutrients to our own cells. Looking outward, we watch rockets roaring through space with liquid oxygen keeping their thrusters alive. Looking around, we see a world where nothing happens unless oxygen is present. Granted, sometimes what is happening is unwanted oxidation such as rust, but that is a small price to pay for air and water and all the rest. Oxygen truly is the life of the party.

When oxygen arrives, he is inviting you to be social. Go out. See your friends. Collaborate on a project. Write letters to pen pals, smile at strangers. Now is not the time to hide yourself away. The world needs you, present and engaged. And set anxiety aside for now. Like oxygen, you may be giving life, or crumbling walls, or even exploding in flames; rest assured that whatever you're doing, it is what needs to happen in this time and place. Laugh and be merry, for you are essential. Now go hug someone.

Fluorine

Element Number:	9
Weight:	18.998403u
Normal State in Nature:	gas
Melting Point:	53.55° Kelvin
Boiling Point:	85.03° Kelvin

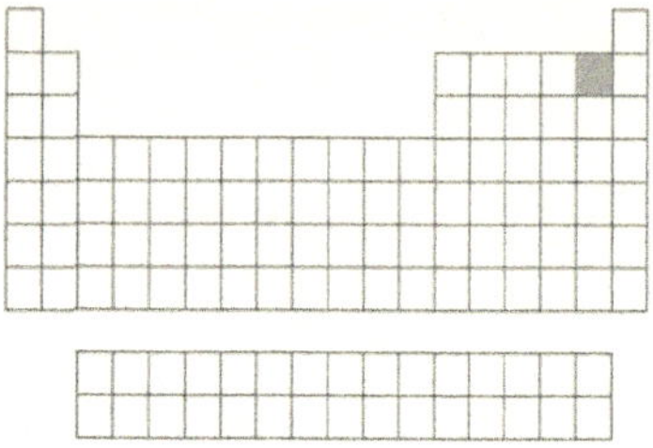

Andre-Marie Ampère is probably the smartest person you've never heard of. Besides pioneering the science of electrodynamics, he also discovered the existence of fluorine as an element in the early 1800s. This was impressive. Fluorine is so reactive, many of the first scientists who tried to isolate it were killed in explosions and dubbed "fluorine martyrs."

Ground-up fluorite is used as an additive which lowers the melting point of metals, and this is where its name comes from, in some languages. Other places call it *ftor*, from the Greek word for destruction. The first alchemist to add ground fluorite to sulfuric acid observed that the resulting substance ate its glass container. Fluorine can be incredibly catastrophic.

There are very few uses for pure fluorine gas. Uranium enrichment is pretty much it. Its compounds, though, are everywhere, from refrigerants to polymers to glass etching. Twenty percent of all pharmaceuticals contain fluorine. It forms bonds with virtually every other element, including xenon.

Calcium, to pick one example, binds fluorine into chunks of hard, white, useful stuff, from which we make toothpaste. It makes the enamel of your teeth hard because calcium grabs the fluoride. It can be found in pesticides, polymers and waterproofing substances.

Oh, and by the way, when not masked by mint flavoring, fluorine tastes terrible.

Anarchy

And by the way, unbeknownst to the new age practitioners who worship them, fluorite crystals actually hate our guts.

-Theodore Gray, *The Elements*

Fluorine is a gas that bursts into flames. That is pretty much its whole repertoire as an isolated element. It reacts with everything except a few of the noble gases, and "react" almost always means "burst into flames." Glass? Flames. Cesium? Flames. Iron wool? Flames. Carbon? Flames flames flames. Like anarchy, you can expect to find fluorine lighting the world on fire. Whether the world likes it or not.

But once the flames are over, it's worth noting that fluorine compounds are incredibly stable. With carbon, for example, it forms Teflon. In nature it is found with calcium in fluorite crystals, which in fact are what the element was named for: *fluo* is Latin for 'flow.' Ground up fluorite is added to metal ores to lower their melting point and make them, yes, flow.

Once the smoke has cleared and the fluorine has bonded, or the anarchy has achieved its objective, be ready for a better flow than there was before. Not to mention stability, and, in Teflon's case, resistance to all kinds of attack. You don't have to like it, and you certainly don't have to trust it, but it is doing what it was made to do. Explode.

RAGE AGAINST THE OPPRESSORS. SET GLASS ON FIRE. BE THE MATCH THAT LIGHTS THE BONFIRE THAT SHEDS LIGHT ALL OVER THE CHANGE YOU WANT TO SEE IN THE WORLD.

*Maybe not literally (unless you want it to be), but now is the time to revolt. Perhaps this means a new haircut and a F*CK CANCER T-shirt, or unfriending some toxic people. Whatever it means to you, rebel! Rattle some cages! Make thrones tremble! NOW! DO IT NOW! BE FLUORINE!!!*

Neon

Element Number:	10
Weight:	20.1797u
Normal State in Nature:	gas
Melting Point:	24.55° Kelvin
Boiling Point:	27.1° Kelvin

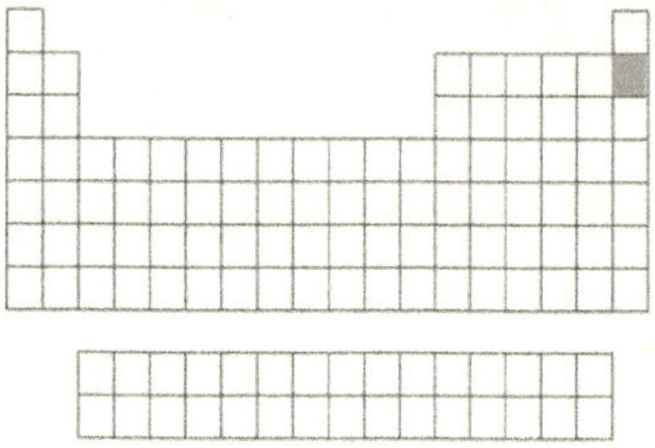

Neon is the second noble gas. Its emission spectrum has a bright red line, and its discoverers named it "neon," meaning "new."

Ramsay and Travers captured some in a sealed tube in 1898, electrified it, and liked the red colored light it gave off. But they were discovering new elements all the time, so to them it was just another pretty face. After all, as far as they could see, it didn't do anything but glow red.

It was introduced to the US market in the 1920s as ultra-modern signage, and scored an instant hit. It remained popular for a good half-century, but was really expensive, and only produced from ores in three countries. Modern lighting uses other gases and other methods, but signs made in the bent-tube style are still called neon.

Neon is created in the universe by stars coming into being, and then it just floats around in space doing nothing. If I wanted an easy career, I'd be neon. It is odorless, colorless and inert. Having a filled outer shell of electrons, it neither needs nor wants to form bonds with any other element. It is common in the universe as a whole, but fairly uncommon here on Earth.

Older applications for neon include wavemeters, lightning arresters on power towers, and cathode ray tubes. It still has applications in cryogenics and lasers, though, so it continues to be cool and cutting-edge. Nearly all the neon produced in the world comes from Ukraine; a reminder that it is important to shine a bright red light on world events.

10

Shine

"Them as has it in them to shine will shine through six layers of muck..."

-Nanny Ogg, *Thief of Time*

Neon does not interact with other elements, nor is it an essential nutrient or particularly important in industry, medicine or agriculture. What it does, when electrified, is *shine*.

The man who discovered this tried at first to market it as an indoor light, but no one was interested in a red-orange hellscape in their living room. One of his apprentices had the bright(!) idea to sell it as eye-catching outdoor signage, which became instantly popular. Times Square and Las Vegas became "awash with neon" overnight. The world got shinier.

All of the noble gases glow when contained by a glass tube with an electric current running through it, but none other has that distinctive reddish orangish lava sunset blood-of-the-gods look. None other is as insistent of a color, demanding the attention of the masses. To be neon is to shine, brightly, unapologetically, and publicly. Neon is garish in private spaces, but oh so perfect everywhere else, the more populous, the better. Shine on.

Drawing neon is your cue to go out and be your most visible self. Introverts can join in too, because one can shine by writing, or creating, or building anything—as long as it will be seen. *That is the important part. Your work, your self, or your vision needs to be public now. Get brave. Put your stuff out there. Give it your best shot. Show them what you're made of.* ***Shine.***

Sodium

Element Number:	11
Weight:	22.989769u
Normal State in Nature:	solid
Melting Point:	371° Kelvin
Boiling Point:	1156° Kelvin

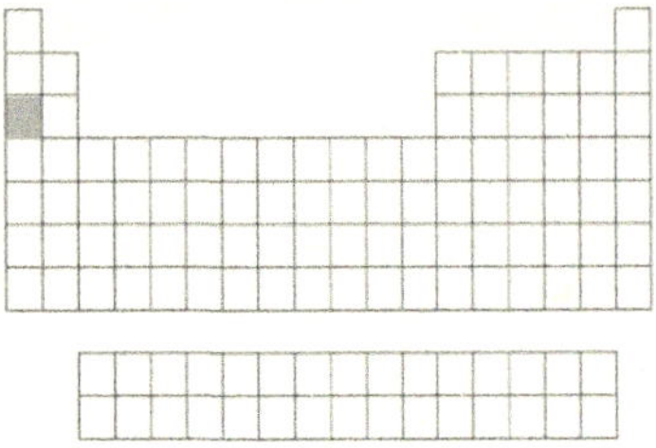

The Roman Empire used to pay its soldiers with salt ("he's worth his salt"), which is where the word 'salary' comes from. The alkali metal itself, however, was first isolated by Sir Humphrey Davies in 1807. He called it "sodium," but Berzelius preferred *natrium*, from the Latin, and that is where the abbreviation comes from. The name sodium is derived from soda, which is from the Arabic *suda*, meaning headache, which it was used to treat. Natrium is from *natron*, an Egyptian mineral.

It is a necessary nutrient in amounts of about 500 milligrams per day, but the so-called Western Diet tends to include quite a bit more than that. The proper amount regulates your blood pressure, while an improper amount dysregulates it. We are what we eat.

Sodium is the second of the alkali metals. Like a lot of the elements, it exists in nature only combined with other things. We are most familiar with it in its partnership with chlorine, as this is the salt in our oceans and on our tables.

If you make elemental sodium at home (DON'T) and any moisture touches it--like you sneeze--it goes boom. Right in your face. It burns with a bright yellow flame, which makes for eye-catching explosions as well as making sodium-vapor streetlights possible. When it's not blowing up or glowing, it gets utilized in glassmaking, soapmaking, paper-making, borax and food.

Quite the useful element.

11

Mischief

Hello, Dad! It is now three in the morning. Do you know where I am?

-Calvin and Hobbes

Sodium is the element of those who aim to misbehave. Whether you want to prank someone with an aquatic explosion or break your diet with a plate of bacon, sodium is your friend. But don't imagine she is no more than a troll; she plays well with others when called for. Together with chlorine she's a major component of the Earth's oceans. And sodium hydroxide is lye, which has served humanity in soapmaking and such for thousands of years.

Sodium is a vital nutrient, but it's quite common to have too much. Particularly in our culture with its high sodium diets, the heart suffers great strain. In similar fashion, we all need a little mischief in our lives, just…not too much. Remember that the world is nearly covered in mischief, that it is a deep, wide, teeming resource and surfing on it is great, but drowning in it…not so much. Do some soap graffiti and run away singing "na na nana na…" And all will be well.

With some inevitable qualms, I must inform you that this card is your carte blanche to commit some mayhem. The kind that involves cherry bombs or spray paint or superglue. Not to make a point or shatter norms, but simply to be a ten-year-old troublemaker. Do just enough to make the world a bit saltier. Don't damage anyone's heart. And don't forget your whoopie cushion.

Magnesium

Element Number:	12
Weight:	24.305u
Normal State in Nature:	solid
Melting Point:	923° Kelvin
Boiling Point:	1380° Kelvin

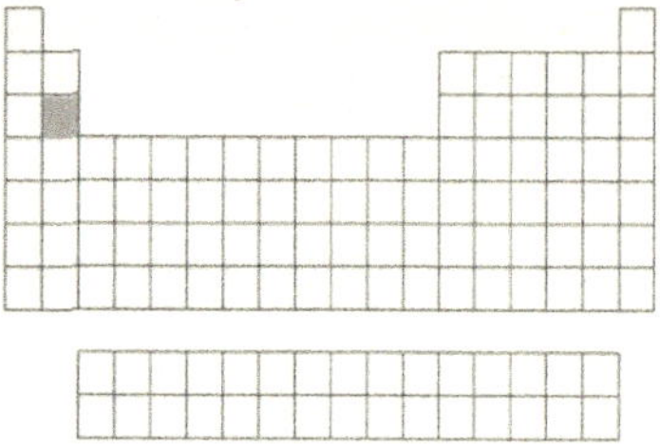

Magnesium was first discovered in Ancient Greece and officially named after the Magnates area of Thessaly. Sir Humphrey Davy got credit for officially identifying it, and its existence was proven by him in 1808.

It is an alkali earth metal with a low density and low melting point. It is made in dying stars and only exists on Earth in compounds and alloys. It is present in sea water and over sixty different minerals.

You need it in your diet to stay alive. All living things use it for enzyme functions and nucleic acid regulation, and in the case of plants, chlorophyll synthesis. It eases sore muscles and gut problems. A severe enough deficiency is dangerous; fortunately, it's a fair bet that at least one magnesium-rich food is something you love to eat.

It burns really hot—like 3,100°C –and you can't put out a fire of pure magnesium. They even make emergency flares with it. Out on the highway, a rainstorm will only make it burn brighter. Its bright white flame is also prized in flash photography and pyrotechnics. Of course, that feature becomes a bug when used in structural components of cars and planes, which has been done many times (and frequently regretted) due to its amazing weight to strength ratio.

Magicians and seers once used it to make fuzzy supernatural beings, ghosts and whatnot. As an alloy with aluminum it is prized for being strong and lightweight, and it doesn't catch fire unless it's powdered. Mostly.

12

Unpredictability

"By-the-bye, what became of the baby?" said the Cat. "I'd nearly forgotten to ask."
"It turned into a pig," Alice answered…

Lewis Carroll, *Alice's Adventures in Wonderland*

Magnesium is a metal that is lightweight and shiny and has hundreds of uses. It's also flammable.

Magnesium flakes burst into a bright white flame and burn quickly to nothing. When used in fireworks or antique flashbulbs, that is a desired effect; when used to build racecars, not so much. How do you know when it's going to burn? You don't! Not for sure. The bigger the magnesium block, the more stable it is, but there is no guarantee.

And then there is its vital nutrient status. Our bodies use dietary magnesium as a laxative or an antacid or even to stabilize blood vessels. Luckily, it exists in many different foods. Even more luckily, if you're magnesium deficient, you can absorb it through your skin. It is super abundant, super useful, super machinable, and all of that comes with the slight inconvenience of never being certain what it's going to do. As we soak in our Epsom salt baths and drink our coffee, I think we can agree that it is worth the risk.

Let magnesium be your inspiration to do things a little differently. Shake up your usual routine. Say Yes when you'd usually say No. Take a fresh route to work, try a new hobby, attack a task you tend to avoid or eat a food you thought you hated. There is a whole world of experience beyond what you're used to, and while you might crash and burn out there, you will also certainly find nourishment for the soul.

Aluminum

Element Number:	13
Weight:	26.981539u
Normal State in Nature:	solid
Melting Point:	933.14° Kelvin
Boiling Point:	2740° Kelvin

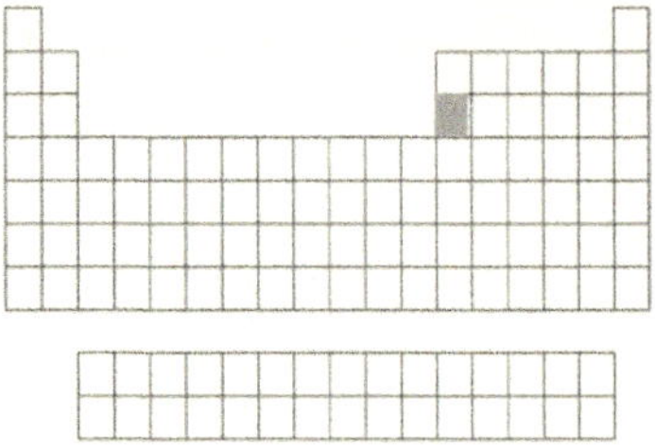

In a paper in 1812, Davies theorized the existence of an element in the substance of alum, which he referred to as *aluminum*. Some of his contemporaries preferred *aluminium*. Both spellings have since been adopted. Ørsted discovered it in 1825, and sixty years later a method was finally developed to mass produce it. It rapidly became very popular and definitely changed the course of history, what with aviation and all.

Aluminum is one of the best known metals in the world. It is used for nearly every commercial product that needs to be metal but also lightweight, safe from most oxidation, and nonmagnetic. If you happen to drink beer, you are already familiar with it. Crack open a can and spend an amusing twenty minutes testing magnets on all the metal objects in your kitchen. If they do not stick, you have likely found aluminum.

Aluminum is about one third the density of steel, and plays well with other metals. Dozens of alloys exist. Your Teflon-coated skillet is not quite the same metal as your kitchen foil. Planes, trains and automobiles all depend on aluminum for its lightness, while canned foods use it because it is nontoxic. Because it transfers heat and conducts electricity, it has hundreds of electronics applications. When it oxidizes, it has this trick of forming a thin but impenetrable layer of tarnish, which then protects the bulk of the metal from oxygen. And, when polished, it is nearly as reflective as silver.

Recycling aluminum takes about 5 percent of the energy it takes to extract it from ores. This fact should get more attention.

13

Renewal

Since I scoured my mind and my body I, too, Lalla, am new, each moment new.

-Lal Ded, *On Her Decision to Stop Wearing Clothes*

Aluminum dances through the world, eternally worthwhile and endlessly renewed. Its lightness makes aircraft possible; its lack of toxicity makes it useful for food and drink cans. It is the go-to material for massive industrial heat sinks as well as sheets of kitchen foil, because it draws heat away from places where heat could do harm.

Unlike plastics, paper or glass, aluminum is easier to recycle than to make from scratch. Like a phoenix, it never really dies; and like a phoenix, it can fly above the clouds on shining wings. But not when it gets bogged down in its old form. The wing casing on a 747 may have been a soda can once, but it would be foolish to think it still is.

Renewal is a concept that gets lost sometimes. We are encouraged to build our resumes, stack our skills, strive forward in our paths but never question them. Perhaps, sometimes, we need to shake free of all that, and decide whether this life really does make us happy. If it doesn't, well…Like aluminum, we have the freedom to change it.

It is time, when aluminum is drawn, for renewal. This can be an ordeal, but trust yourself. Examine old habits and assumptions. Become keenly aware of negative beliefs you don't need anymore. You can even have yourself a good meltdown, and while you're molten and incandescent, choose a new form. Start fresh. Become yourself, only more so. Throw away the dross that has been holding you in an obsolete shape. Keep those traits that are of real value to you, such as your abilities to step lightly, move heat around, and resist corrosion.

Silicon

Element Number:	14
Weight:	28.0855u
Normal State in Nature:	solid
Melting Point:	1683° Kelvin
Boiling Point:	2630° Kelvin

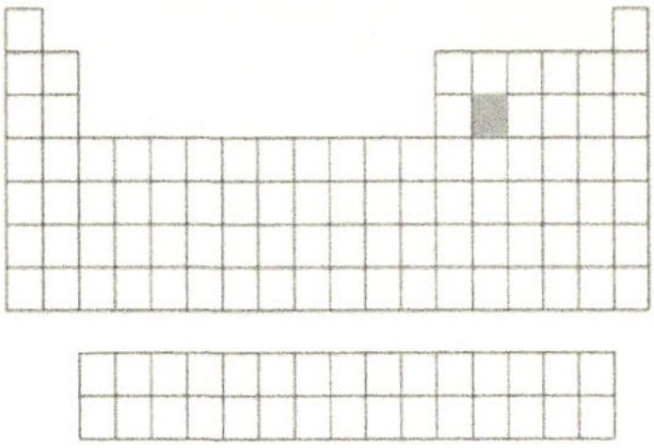

Silicon is a metalloid, which means it behaves like a metal in some situations but not in others. When bonded with oxygen (which it loves to do), it becomes a silicate, and silicates make up 90 percent of the Earth's crust. The most obvious place to look for silicon, therefore, is in a rock collection or mineral museum. However, you will also get lucky if you look in your phone. Most electronics these days utilize silicon's semi-conductor traits and microscopically flat surface.

Its bond with oxygen is so strong, it was very difficult to isolate, even when scientists were certain it existed. Officially, it was discovered by Berzelius in 1824. Most languages call it variations on Davy's *silicium*, named for the Latin word for "flint," but in English we use Thomson's *silicon* (his logic being that *-ium* is for metals, which silicon is not).

There is an ongoing argument about whether silicon is an essential nutrient. This is hard to measure because of how difficult it would be to achieve a deficiency; our planet is mostly silicon and we all eat our daily peck of dirt, one way or another. It definitely does improve our bone density. Plants use it to strengthen cell walls. Some sea creatures even use it to build their entire skeletons.

As a free element, it's rare. It loves things it can make salts with, which is almost everything. Sand and mortar, glass and ceramics, clay and cast iron, those little DO-NOT-EAT humidity-absorbing packets. Even modern cutting boards and spatulas use it, because the melting point is so high, and it doesn't trash your knife blades.

14

Order

Everything that happens within the created Order, no matter how small, has its effect.

-Madeleine L'Engle, *A Swiftly Tilting Planet*

All of the rocks, sand, mortar and mountains you've ever seen contain silicon as a main ingredient. It lends its structure to glass, ceramics, and of course those rubbery silicone polymers that borrow the name. But it's when we look closer that we see how frighteningly coordinated this element is. Silicon is smooth enough to host computer circuitry which is literally microscopic. Mountains are impressive, but the computing power we have in our cellphones is mind-boggling, and both of these are dependent on silicon.

Order is key. As above, so below; as the mountains are strong, so the motherboard of your laptop is durable. The mortar that binds your foundation together is like to the artificial mind keeping track of your appointments. Without it, there would be chaos. Like its older sister, carbon, silicon has the capacity to form complex connections, and somewhere in the universe there may well be silicon-based life. If that is so, we can be assured that they are an integral part of the order of creation.

When Silicon is drawn, it is counselling you to pay attention to the order of all. Whether you meditate on the entire universe, the fate of the planet or your own personal circumstances, try to step back and see the harmony between elements, the unfolding of patterns and the interwoven nature of events. From here, if you observe discord, you can do something about it. Neglect neither the large, mountainous pictures nor the tiny, complicated details. Order is in the whole.

Phosphorus

Element Number:	15
Weight:	30.973762u
Normal State in Nature:	solid
Melting Point:	white: 317.3° Kelvin/red: 860° Kelvin
Boiling Point:	white: 553.7° Kelvin

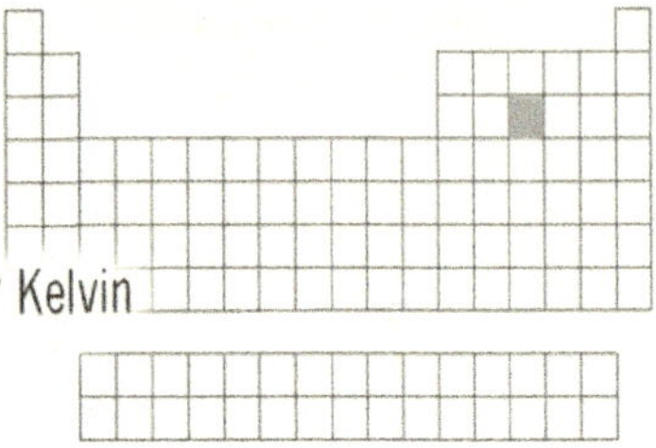

When an element can form several crystal structures, the macroscopic traits of that element vary widely. These different forms are called *allotropes*, and most of the elements between group 13 and group 16 have multiple allotropes. Phosphorus is no exception.

White phosphorus is insane. It's the only element that is a Schedule 1 substance for the DEA; it glows greenish in the presence of oxygen (hence "phosphorescence"), and is very, very pyrophoric. Given the tiniest excuse, it will burst into flames that are nearly impossible to extinguish. Its most notorious application is in warfare, where it teams up with napalm to spread unspeakable misery and pain. Its name comes from the Greek *light-bringer*, which in Latin would be *lucifer*, and there is poetic accuracy in that.

Black phosphorus just sort of exists.

Then there's red phosphorus. In this form, phosphorus is an essential nutrient for every form of life, especially plants. It's the P in NPK fertilizer, and your own body uses it to support the structure of your DNA. The transportation of cellular energy would be impossible without it. As a point of interest, red phosphorus does not boil, but sublimes between 689° and 863° Kelvin. But on a more practical level, it is often the limiting factor in how much life a given patch of soil can support; and if that's not enough, it is also the main ingredient in matches.

Always with the burning.

15

Prosperity

But there was always enough to eat, and the water always flowed in hot months, while a fire always burned during the cold ones.

-Grace Lin, *Where the Mountain Meets the Moon*

What does prosperity mean to you? Plentiful food? Red phosphorus fertilizer can help you grow it. Is it warmth and light? Strike a phosphorus-tipped match and use it to start a blaze in your fireplace. Is it the power to hoard wealth and incinerate anyone who complains? Well, unfortunately, the white phosphorus allotrope can help you there. Perhaps you want this because you believe that prosperity is a zero-sum equation; that for you to be content, someone else must suffer. But if that is your goal, you've already missed the point of what prosperity should be, as well as forcing blood on the hands of an innocent element.

Like love, kindness or economic stability, prosperity grows when it is shared. Let no one stay hungry, let no one live in the dark or the cold. Follow the lead of the red phosphorus, and heed the presence of the white only as a warning. Because at its root, prosperity is not unlimited wealth or waste, excess or greed; prosperity is knowing that we have what we need. And once we know that, we can thrive.

When this card appears, it is drawing your attention to your sense of security and general welfare. Are you, by your own definition, prosperous? If not, what can be done to change that? It is a truth rarely accepted that whether we feel lack or abundance has less to do with our conditions and more to do with our conditioning. A rich man can fear lack; a student drowning in debt can feel rich. Do you have food? Warmth? Light? That's the place to start. If you do, feel glad. If you do not, let phosphorus help you find them.

Sulfur

Element Number:	16
Weight:	32.065u
Normal State in Nature:	solid
Melting Point:	392.2° Kelvin.
Boiling Point:	717.8° Kelvin

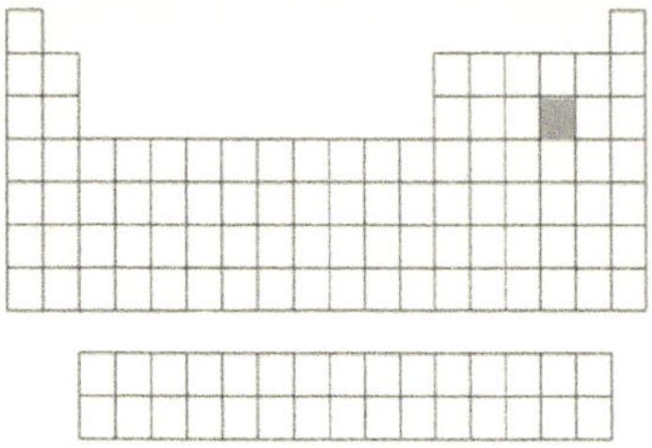

Sulfur never needed to be officially discovered. It is probably the oldest element known to humans. It isn't even named after anything, but has always been a word in its own right--although some confusion happened between the Greek and the Latin to give us the modern word.

It was already unspeakably old when the Old Testament was written. It was used in ancient Greece to fumigate houses, in ancient China to make black powder, in ancient Egypt to treat granular eyelids and in ancient Indian alchemical experiments. It was particularly easy to use because its native form can be found, not just its oxides.

Like its neighbor phosphorus, it burns easily, and the smoke is smelly. It's used in wooden matches, explosives, and fireworks.

Rotten things give off sulfur gases. In fact, most of the unpleasant smells you have smelled are organosulfur compounds. But do not be fooled into thinking that it's only an ugly face. Without it in your body, you die. It's part of your DNA and stabilizes your proteins.

A lot of American sulfur comes from refining petroleum. And it's used for everything. It shows up in all kinds of pharmaceuticals, from acne cream to antibiotics; it will heal parasites, worms, dermatitis, fungal infections and a plague of locusts. It is considered an organic insecticide. Plus it fertilizes fields. It's just good for a lot of things. But very, very stinky.

16

Medicine

This project is my medicine.

-Laura Kampf

Sulfur has been with humanity since the very beginning. It is mentioned in holy books and ancient writings from Egypt to China; some scholars back then called it brimstone. And everyone had two observations to make: 1. It is good medicine for conditions ranging from scabies to constipation; and 2. It smells bloody awful.

Sulfuric compounds are the reason your hair and nails stink when they get singed. They are the signature smell of burnt eggs and hot springs and flatulence. When a medicine tastes terrible, chances are high that it contains sulfur. And chances are also high that it will work. Medicine is not supposed to be *nice.*

Even more: several amino acids and vitamins are organosulfur compounds. At the cellular level, life literally cannot exist without it. But where life has become parasitic, sulfur is your medicine. A blight on your crops? Eczema on your elbows? Let there be brimstone. Even winemaking requires it, to keep harmful bacteria at bay while the good bacteria does its fermenting. Sulfur is medicine for all.

When you have drawn sulfur, it is time to prepare the way for healing. What are your parasites? Want to lose them? Using sulfur's stanky energy will never be pleasant, but it will free you from illness of the mind and slough off the mildew of the soul. Now is the time. Look hard at yourself, at the blights you've learned to live with, at the places where wellness just doesn't come. Think of the unpleasant but necessary tasks you must tackle in order to change those realities. Now, hold your nose and take your medicine. It's going to get better.

Chlorine

Element Number:	17
Weight:	35.453u
Normal State in Nature:	gas
Melting Point:	171.6° Kelvin
Boiling Point:	239.11° Kelvin

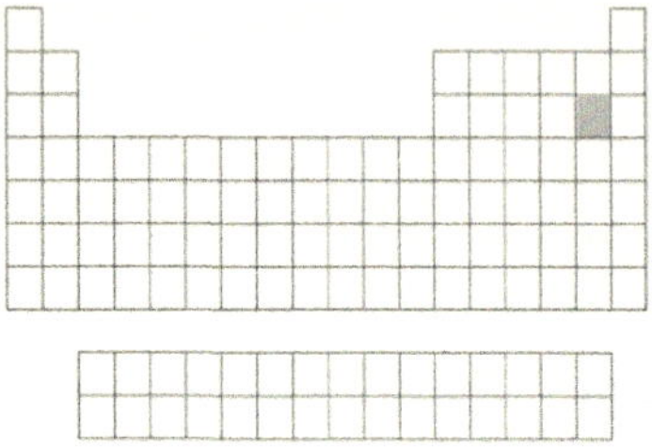

Chlorine is a halogen that lives just under fluorine on the table. It may not be quite as insane as its sibling, but is still extremely reactive and corrosive. Alone, it is a yellowish-green noxious gas, but it is very rarely alone, being much more comfortable in compounds. There are more than fifteen thousand chlorine compounds. The Earth's largest deposits of it are in seawater. It's been used since 6000 B.C., and was a fan favorite with medieval alchemists. It will even dissolve gold when mixed with nitric acid.

In 1630, van Helmont recognized it as a gaseous substance, not just a component of so many liquid solutions; and in 1810 Davy confirmed that this gas was, indeed, an element. He named it using the Greek word for pale green. The solutions still have more uses than the gas itself, as antiseptics are welcome in so many places.

Chlorine is yet another element that will kill you if you don't get enough. It is an essential nutrient for metabolic processes, and a main ingredient in your stomach acid. Fortunately, it's everywhere. Mix it with sodium and you have a dozen different kinds of salt. It was a germicide before anyone knew what germs were, and is the reason you can trust the water from your kitchen tap.

But, of course, anything this corrosive would have to be weaponized. It was used to melt soldiers' lungs in WWI and is still occasionally used in truck bombs in the Middle East. Because it is impossible to aim, though, it has never been anyone's favorite mass murderer. For the most part, people would rather use it to disinfect than to destroy.

17

Choice

And just when you'd think they were more malignant than ever Hell could be, they could occasionally show more grace than Heaven ever dreamed of. Often the same individual was involved. It was this free-will thing, of course. It was a bugger.

-Neil Gaiman and Terry Pratchett, *Good Omens*

Chlorine gas was used in trench warfare in WWI to kill and maim tens of thousands of soldiers. It was indiscriminate; the side setting the canisters was as likely to die as the opposing men. A shift in the breeze, and everyone's lungs would fry. It was appalling.

And then again, small quantities of chlorine will make questionable water fit to drink. Millions have been saved from slow, disease-ridden death by this miracle element. It is a boon to mankind, the substance of hope to isolated villages, and the pleasant smell of a swimming pool on a bright summer morning. It is wonderful.

Any of us has the capacity to harm others in irreparable ways. We also have the power to protect those same others, shielding them from that which would destroy them. Everything we do has an effect on our world and our people. The difference between great evil and great good is not predetermined. It lies in the choices we make.

The lesson chlorine has for you today is: You are not a neutral party. You cannot be. Your actions and inactions have consequences that ripple outward and change the world. That is a heavy load to bear, but it is unavoidable. No one can make the decisions for you, and no one can make your decision not count. So, which is it to be? Will you be an agent of destruction, or a champion of protection? The choice is yours.

Argon

Element Number:	18
Weight:	39.948u
Normal State in Nature:	gas
Melting Point:	83.95° Kelvin
Boiling Point:	87.302° Kelvin

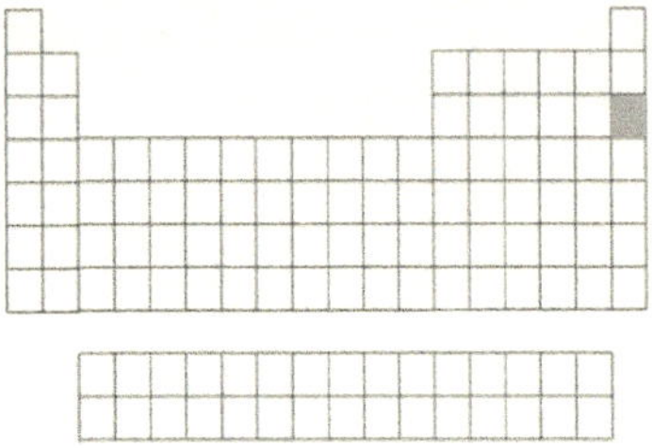

Argon is by far the most abundant noble gas in the Earth's atmosphere, making up nearly one percent of the air we breathe. It exists in the crust, too, as a result of the decay of radioactive potassium. It is everywhere, which makes it surprising that no one identified it until the late nineteenth century. It was the first of the noble gases to be isolated, by Rayleigh and Ramsay in 1894. They were recreating an experiment that Cavendish had come up with a hundred years earlier.

But in their case, it was successful.

Most argon in the universe is made by cosmic rays, or supernova nucleosynthesis if you want to get fancy. Argon lasers have many medical applications and argon-filled glass tube lighting glows pale blue.

When did you last use argon? When you opened a bag of chips or salad? Or perhaps when you did a little light welding? Argon is used in both of these situations, as it doesn't play well with others. It pushes other gases out of a space, preventing oxidation. This also makes it useful in fire suppression, particularly where there are expensive electronics. Just don't be in the server room while this is happening, as argon will sink to the ground and asphyxiate you as you are trying to crawl to safety.

It's a doomed-either-way situation.

Argon was named for the Greek word for "lazy," which is a harsh way to treat the gas that keeps your chips fresh and your welds clean.

18

Stillness

To the mind that is still, the whole universe surrenders.

-Lao Tzu

Argon glows sky blue when an electrical current wakes it up. Otherwise, it sleeps.

It does not wake up—that is, react—when subjected to heat, or exposed to other elements. When oxygen wants to play, it sleeps. When fluorine would cause a ruckus, it sleeps. Argon is used as a shield gas for any operation where a chemical reaction is undesired, because argon is plentiful, cheap and totally inactive.

Stillness can be found in unexpected places, at quixotic times. Just as one can benefit from taking a moment to breathe in a bustling airport, or grab a catnap after lunch on a busy work day, so can argon be found using its inactivity to allow things to happen. Whether you're welding steel or preserving your open bottle of wine, this sleepy fellow is your friend.

Take a breath.

Take another.

We forget, sometimes, how necessary stillness is. We live in a world that practically worships work, where "productivity" is used as a synonym of "worth." Forget all that. Argon is telling you to rest, right now; no phone, no news, no screens, no interruptions. Find a sunbeam, and a cup of tea, and rest.

Potassium

Element Number:	19
Weight:	39.0983u
Normal State in Nature:	solid
Melting Point:	336.7° Kelvin
Boiling Point:	1030.79° Kelvin

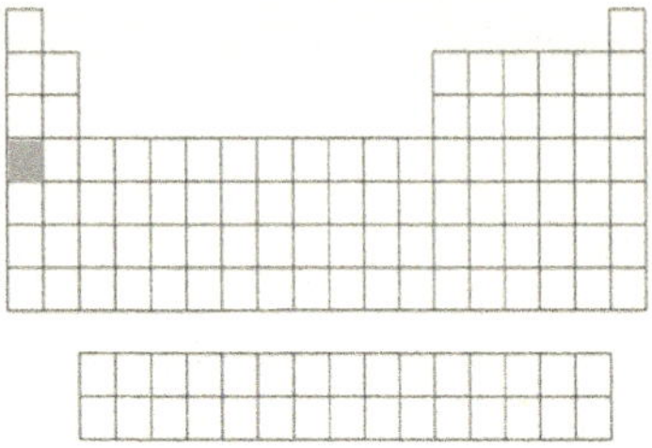

Potassium was isolated by Davy in 1807, using electrolysis. But the potash for which it was named has been a staple of human civilization since forever. Glass and soap, gunpowder and bleach, all use it. And the list is quite a bit longer than that.

Its other name, *kalium*, is derived from *kali*, the Arabic root of the word *alkali*, which also simply means plant ashes. Gilbert and Berzelius preferred this name, which is why the symbol for the element is K.

Like the other group one members, potassium doesn't especially like being a raw alkali metal. It has an electron floating by itself, looking for a connection, and there are lots of willing suitors. Especially when it connects with water. The heat makes the hydrogen, released by the water, burst into bright magenta flames.

Your body uses potassium to control your heartbeat. It also uses it to regulate hormones, electrolyte balance, blood pressure, and gastrointestinal motility. It's a Swiss army nutrient, in fact, so a deficiency is taken very seriously. If you've ever tasted one of those potassium pills, you know it's nasty. But worthwhile.

Plants need it too, and the preferred fertilizer for a lot of crops is potash, which is a potassium salt. Country dwellers get used to seeing the trailer with the tank on top driving through the fields. Over 90% of the world's commercial potassium production goes to agriculture, as it is one of the three macronutrients plants need. It is the K in NPK. It feeds us all.

19

Nurturing

"All nice things are good for you."

-Moominmama, *Exploits of Moominpappa*

Potassium is the elemental matriarch. It nourishes and sustains, not just us, but plants as well. Ancient cultures used it, but they didn't know it was an element; they just knew that, in the form of potash (which the Romans called Kalium), it was vital for gardens and soap. Nowadays we think of it as a substance you find in bananas, which is true. It's also in mushrooms, melons, avocados and sweet potatoes. And so much more. Nurturing is also found in more places than you'd expect, and might come from sources you hadn't noticed. Which is fine; one of the hallmarks of those who nurture is that they really don't give a damn about being recognized. Potassium's role in the human body is crucial. Without it, muscles freeze and nerves go awry. A severe enough lack will even stop the heart. A lack of nurturing can have the same effect. Be wary of the lie that a person can get along without it.

But there is another side to this vital, reassuring element, and that is the fact that it is also an alkali metal, which means it will explode violently when dropped into water. Violet-ly, too! The flames it gives off are a lovely reddish purple. This may seem unrelated to nurturing, but sometimes we just need to be startled into action. And the person who knows when we need a hug is always the person who knows when we need a kick in the butt.

Look at Potassium! See her little face! She's saying, "Who needs some banana bread? Or a nice soapy bath? Or a well-tended garden? Or to be startled out of their ennui by explosive purple flaming rocks from hell? Got you covered, dearie."

Now decide whether you need to have that person in your life right now, or BE that person for someone right now. Or maybe you need to be that person for yourself. Whichever way, nurturing needs to happen.

Calcium

Element Number:	20
Weight:	40.078u
Normal State in Nature:	solid
Melting Point:	1112° Kelvin
Boiling Point:	1757° Kelvin

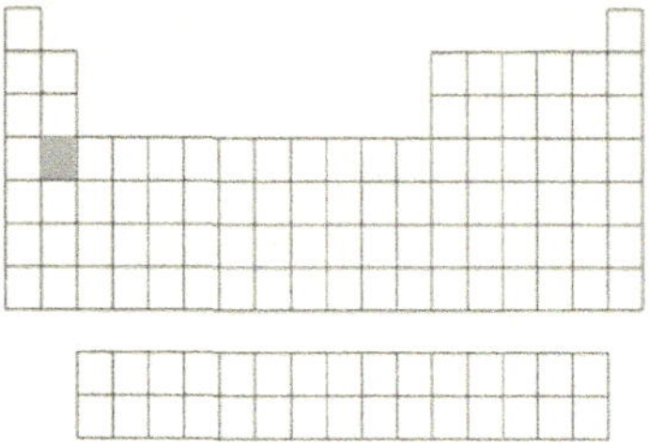

Lime and gypsum been used by humans since forever, but were not suspected of being the oxides of an actual chemical element until Lavoisier surmised as much in 1789. Davy proved him right in 1808 when he isolated the metal itself.

Calcium is a strong regulator of your heart and blood pressure. Too much or too little are both a problem. It's the most common metal in your body, and is used for everything from neurotransmission to protein synthesis. It pairs with vitamin D to make and harden bones. Bones are largely hydrated calcium phosphate, and some speculate that the skeleton's role as a structural necessity for vertebrates is actually secondary to its role as a calcium reserve; this may sound bizarre, but is supported by the fact that a body which is struggling to find usable calcium will leach it from its own bones.

Calcium is an alkali earth metal, but we rarely talk about it as a metal. Virtually all its applications are salts. It's used in construction, cement work, pottery, and nutrition. It improves the castability of steel, the strength of aluminum and the durability of cast iron.

Calcium salts are nutritous and benign, but calcium alone can be scary. It can be found in car batteries and drain cleaner, but that's about it. Pretty sparce compared to the thousands of critical uses for its compounds.

Calcium deposits in the form of calcium carbonate are all over the world, the remnants of ancient sea life turned into a soft white rock. It makes its way into food, which is lucky for us. Best not to let your body sacrifice your bones.

20

Structure

"I make all good things."

-Jimmy Diresta

Most of us, when we think of calcium, see in our minds bones, cliffs, eggs…Things that are rigid and dependable. Structures. All of these things are compounds, of course, because calcium itself is an alkali earth metal and therefore soft and shiny in its pure form. But its pure form oxidizes quickly, and is not terribly impressive. Once it has joined forces with carbon or sulfur or phosphorus, however, it becomes a structural force to be reckoned with.

The concept of structure has similar properties. We cannot build anything worthwhile by simply aiming at rigidity for its own sake. But if we apply structure to our endeavors in Life, or Prosperity, or Medicine, suddenly we find a fundamental power greater than the sum of its parts. A coral reef is more than calcium carbonate; your skeleton is more than hydrated calcium phosphate. Everything that has structure is more than it appears at first glance. The beauty of a tiny, twisted seashell and the awe inspired by the White Cliffs of Dover depend on this one metal, and even your cells require it, to mediate and give structure to various processes. Metaphorically and literally, calcium gives the world backbone.

Drink your milk and stiffen your spine.

Calcium is telling you that something in your world needs more structure. Looking honestly at your life, do you see anything that is being ineffectively executed, or even neglected? You do not need to suffer any heartburnings or guilt about it, just try to address it. Put it in your routine. Make it part of your rhythm. A schedule is a dance, not a prison. Choreography is your friend. And don't get too hung up on routine for its own sake. Remember, calcium does its best work when it is part of a greater whole. And so do we.

Scandium

Element Number:	21
Weight:	44.955912u
Normal State in Nature:	solid
Melting Point:	1814° Kelvin
Boiling Point:	3109° Kelvin

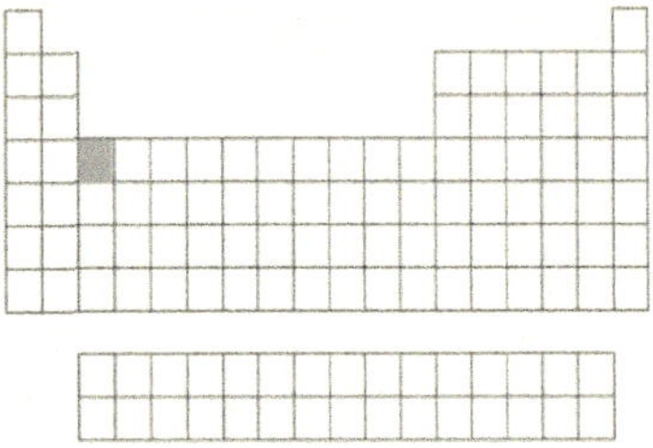

Named for Scandinavia, Scandium was discovered by Nilson in 1879. Dmitri Mendeleev had predicted its existence and some of its traits ten years earlier, but Nilson did not know that. It was a coincidence.

There are no significant deposits of scandium, despite being more abundant than either lead or mercury. It is simply dusted throughout the Earth's crust. Like powdered sugar.

Scandium is classified with the rare earth elements, as it tends to be present in the same ores as yttrium and the lanthanides. However, it is more correct to call it a metal, one which is not actually rare but also is not concentrated in any location, rendering it very expensive to gather and refine. Most of the small amount that is produced annually comes out of China.

The demand for it is consistently higher than its production, because it combines with aluminum to create a very strong, very lightweight, corrosion-resistant alloy used to make the highest quality baseball bats and bicycle frames, and even some small airplane components.

Scandium-based halide lighting is utilized in places where a day-like light temperature is desired, such as sports stadiums. Because it is resistant to softening at high temperature, it has yet another application in lithium ion batteries.

But that's pretty much it. It would probably be used in more industries if it could be gathered and refined with less expense. Until then, its list is short.

21

Confidence

I consider it a challenge before the whole human race
And I ain't gonna lose

-Queen, *We Are the Champions*

Scandium is considered a rare earth element, although the words "spread thin" would be more appropriate than "rare." It is difficult to collect and refine, and is therefore expensive. Confidence can also be thin on the ground, and time-consuming to gather. But once that is accomplished, scandium is hard, strong, and excellent for alloys in all kinds of applications. Really, the only downside to this element is its expense. Sometimes it costs us a lot to have confidence, and that price should not be ignored.

But it's always worth it. If you play baseball or do competitive cycling, you will find confidence--and scandium--essential. It's even used in airplane parts, because of its resistance to oxidation. And there's no better feeling than confidence in your airplane parts.

Also, if we want our stadiums or parking lots brightly illuminated with a clear white light, scandium halide lighting is the thing we need. When we show confidence to the world, we shine brightly and can shed our effulgence on places and things that must not be left in the dark. Light, bright, strong, and rare…why shouldn't we be confident?

Scandium is here to defeat your self-doubt. At this time and place, the voices in your head that wonder if you're on the right path, doing the right thing, among the right people…are horse crap. Ignore them. You are exactly where you need to be, doing what needs to be done and can be done by nobody but you. Square your shoulders and stride forth proudly. Confidence is yours, it belongs to you, you embody it and you deserve it.

Titanium

Element Number:	22
Weight:	47.867u
Normal State in Nature:	solid
Melting Point:	1945° Kelvin
Boiling Point:	3560° Kelvin

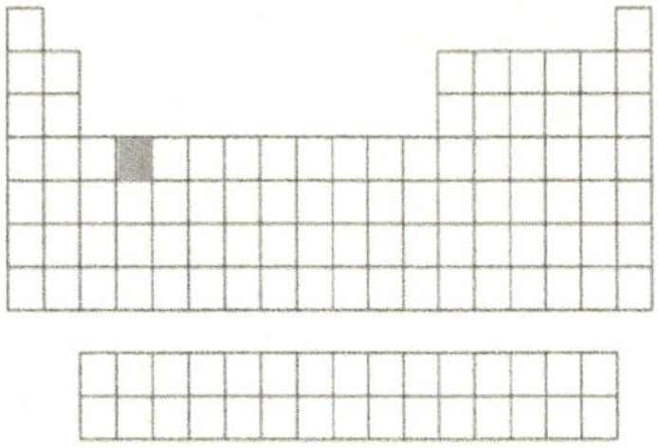

An English geologist named Gregor was the first to notice a strange oxide in a handful of sand in 1791. Four years later, a German chemist named Klaproth independently isolated the metal and named it after the Titans of Greek mythology.

Titanium is found in nature as an oxide, and it takes some work to reduce it to its metallic state. Once achieved, however, titanium is one of the strongest, lightest transition metals, and one of the most resistant to corrosion. It alloys well with other metals, bringing its traits to applications ranging from dental implants to aerospace components.

It has a comparatively high melting point, which makes it a useful crucible for smelting other metals. It's not terribly rare, being the ninth most abundant element in the Earth's crust, but can be expensive because of its high demand.

Only about 5% of titanium-containing ores are refined into its metallic form; the rest are destined to become pigment. Painters will recognize titanium dioxide as the pigment that makes the brightest white. It is an ingredient in artificial smoke and the manufacture of polypropylene products.

The moon rocks brought back by the Apollo 17 mission were over 12% titanium.

Because it is inert in bodies, it is the go-to metal for knee and hip replacements, and also surgical instruments and expensive sports equipment. Some of the products that claim they are titanium do not, in fact, contain any at all; the word is used for marketing purposes because it is widely recognized. Many people associate titanium with excellence.

22

Imperviousness

Don't let the sand blind your eyes
Realize they're only lies and
Keep on walking

-Grace Potter, *Oasis*

Titanium is one of the more famous of the rough, tough transition metals, but its fame does not affect its performance. Not much does. It is resistant to oxidation, to chemicals, and to time. Just in general, titanium is impervious to corruption, or change. It is low-density, very strong and…heedless.

Because there are two sides to imperviousness as a personal quality. Incorruptibility is praiseworthy, but insensitivity is less so. Both of these are titanium-like traits. One who refuses to compromise their ideals may well have difficulty in compromising at all. Not that titanium plays poorly with others; it is frequently alloyed with vanadium, iron or aluminum, to name a few, and the results are always lightweight and strong enough even for aerospace projects. Not to mention golf clubs, aircraft recorders, tools, prosthetics and watches. But can there be any higher calling for a metal than space itself?

Titanium's imperviousness puts the stars within our reach, and for that, its lack of tact is forgivable. Slap some white paint on any gaffe it makes, and move on.

Things may try and mess you about, but with titanium in your spread, the message is to hold firm to the ideals you know to be true and right. Do not let anything cause you to compromise what you hold most dear. Stay strong and incorruptible.

Pay attention, though: titanium may be giving you a gentle warning about being oblivious. When your line in the sand destroys someone's sandcastle, you must first notice that you have caused harm—which can be a challenge—and then make amends. This can be done without losing your strength of purpose.

Imperviousness is your byword.

Vanadium

Element Number:	23
Weight:	50.9415u
Normal State in Nature:	solid
Melting Point:	2163° Kelvin
Boiling Point:	3650° Kelvin

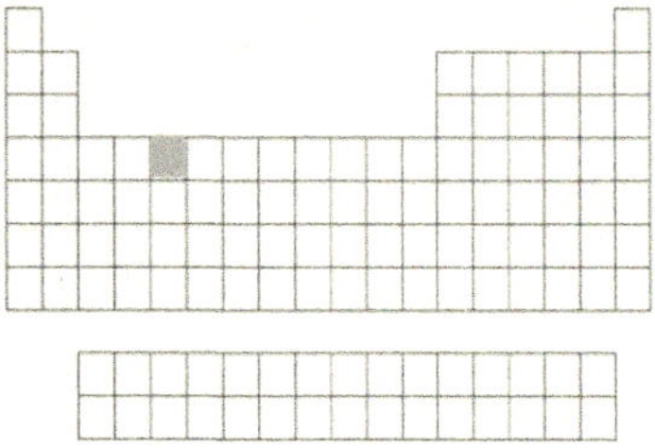

Del Rio, the scientist who first found a mineral rich in vanadium, was sure that he'd found a new element...but second-guessed himself and did not publish his findings. This was in 1801. In 1830 Serfström managed to create a chloride of vanadium, proving that it was a new element after all, and he named it after Vanadis, the Norse goddess of beauty and fertility. His stated reason for using that name was that vanadium compounds contain many lovely colors... and besides, there weren't yet any elements beginning with V.

Vanadium is difficult to isolate, so it was 1867 before the pure metal was finally made available. It turned out to be ductile and malleable, resisitant to corrosion and, while not terribly impressive by itself, its alloys are stronger than titanium.

85% of the vanadium produced annually goes to those metal alloys.

The first major industrial use was to make the chassis of the Model T. It is crucial for high-speed tools, crankshafts, and basically any machine component where strength is paramount.

It is being considered for use as a high energy density anode in lithium ion batteries.

In large amounts, all salts and compounds are toxic to humans. Vanadium occurs in combination with uranium, which was mined in western Colorado for years. China is now the world's largest producer.

Some sea creatures and mushrooms use vanadium as an essential nutrient.

23

Strength

Don't let it tear you down. Let it strengthen you up. I can only fight because I've learned to.

-Bernadette, *The Adventures of Priscilla, Queen of the Desert*

Strength is one of the most mythologized virtues. From the very beginning of humanity, our stories have been about people who were strong. Beowulf, Durga, Gilgamesh…where there is strength, there are heroes. We look for that in ourselves as well. What can we weather? How much can we endure? As tools in the hands of our own destinies, are we strong enough to do the job we came to do?

We all want to be heroes.

Vanadium is that kind of hero. She is not nearly as famous as titanium or iron, but she is much, much stronger. Vanadium alloy tools, drill bits, and machine parts are an absolute necessity for those who know the value of stress-resistant, heat-resistant, wear-resistant strength. In a community of makers or builders, vanadium is recognized as essential. In other company, her name is unknown. But then, so is Gilgamesh. Why worry about what the ignorant think? Those who count, know.

By drawing vanadium, you are also drawing strength. Did you doubt your ability to get through this time? Not only will you get through it, but you will do it with a grace that could be called heroic. Your strength is legendary, unassailable, unstoppable. Perhaps not many people will see what you have achieved; but those who matter will not just see, but applaud you. Your strength will inspire your tribe.

Chromium

Element Number:	24
Weight:	51.9961u
Normal State in Nature:	solid
Melting Point:	2130° Kelvin
Boiling Point:	2945° Kelvin

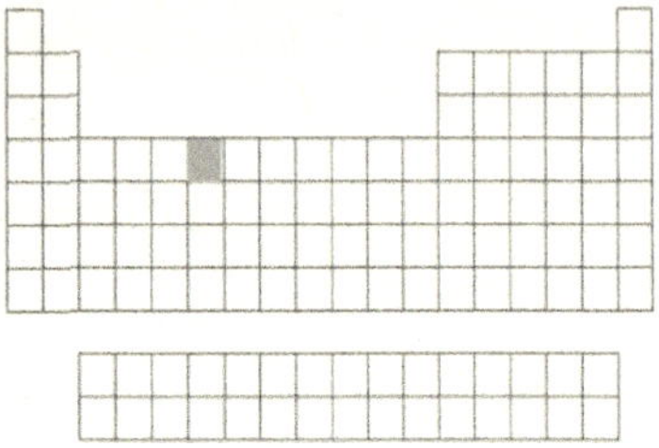

A geologist and a zoologist were the first to find chromium-rich minerals in the 1760s, but they simply used them in pigments without really getting to the root of what made those minerals colorful. It took over thirty years for Vauquelin, a chemist, to get his hands on a sample, and he was able to isolate the metal and is therefore credited with discovering it.

It was named chromium because so many of its compounds are vividly colored. Rubies, garnets, sapphires and emeralds all contain chromium, but the hues don't stop there. Paired with lead in a mineral called crocoite, it becomes a dazzling crimson. Its compound with sulfur is sunshine yellow, and combined with chlorine it becomes a deep, royal purple.

So artists are familiar with chromium, in its rainbow of forms. It has also been used to tan leather. It's a necessary component of stainless steel, being tough and corrosion-resistant. Its high index of refraction makes it popular for shiny, durable finishes on a myriad of surfaces. American automobiles sported chrome clad trim and bumpers, starting in the 1930s. It is still common on motorcycles. And bathroom faucets.

Biologically, chromium is a key player in blood sugar control, but some forms are extremely toxic. Until the toxicity became a concern, school buses were painted with chrome yellow. It's still used by the military, because the green pigment reflects infrared light to the same concentration as green leaves.

24

Swagger

Larger than life…Is just the right size.

-Noxzema Jackson, *To Wong Fu, Thanks for Everything, Julie Newmar*

Chromium is a key component in stainless steel, and it also pairs up with vanadium to make some really high-performance tools. It is a fine, corrosion-resistant workhorse of a metal; but when the serious jobs knock off for the weekend, that's when chromium lets its hair down.

Because Chromium is *shiny*! Its minerals are technicolor and its metal is lustrous. It reflects and gleams and defines curves. It lights up boring surfaces and makes even the bumper of a Buick look like a party at the club. Chromium is at home in Vegas, in Reno, at car clubs and in places where "tone it down" are words you just don't hear. Most chrome surfaces really are only surfaces; a microscopic layer of chromium can be plated onto anything to give it that gorgeous—or tacky, depending on who you are--sheen. But when we start to equate that with shallowness or weakness, we would do well to think of chromium's crucial role in stainless steel. Sometimes swagger on the surface is only hiding muscle underneath.

Break out your purple platform stilettos and get your ass in that spotlight. Swagger is more than boldness, more than audacity, more than glamour. Swagger is all of these mashed together and covered in sequins. Whatever you're doing, make it over-the-top. Make it memorable. Make it loud.

Give it some swagger.

Manganese

Element Number:	25
Weight:	54.938044u
Normal State in Nature:	solid
Melting Point:	1518° Kelvin
Boiling Point:	2335° Kelvin

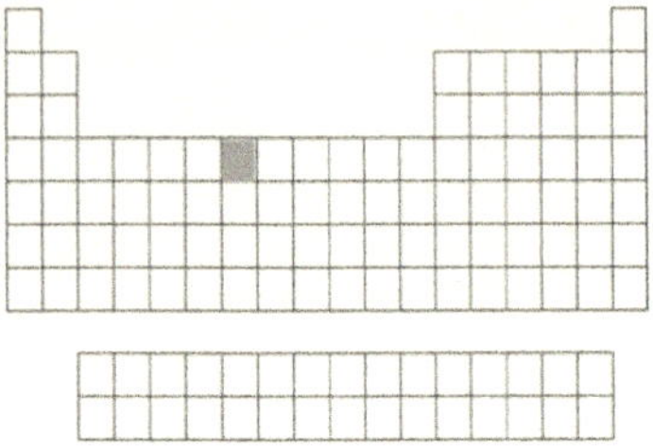

Named for the Magnetes area in Greece (sort of. It's complicated), manganese has been around for a very long time. It is a black pigment in paints, and has been found in cave paintings that are 30,000 years old. It has been used by glassmakers for hundreds of years because it can color the glass, or remove the color from glass. One salt was called glassmaker soap.

As can happen when something is ubiquitous, it had a thousand uses before anyone wondered what it actually *was*. Gahn was the first to isolate a sample of mostly pure manganese in 1774, and it was found to be a metal; one that can improve the quality of steel or aluminum alloys by removing sulfur, phosphorus and oxygen impurities. It does this by absorbing them.

Manganese is used to protect gun barrels from rusting. It's an ingredient in batteries, and was used in place of nickel in coins during World War II. Its alloy with copper is used in shunt resistors in automotive and industrial settings, because it is largely unaffected by sulfur compounds or temperature changes.

In medicine, permanganates are common for dyeing lab samples.

Manganese is also an essential part of your diet. It is important for dozens of enzymic functions and is also an antioxidant. Just like in metal alloys, it removes free radicals that would otherwise do damage.

Don't worry, though. It is very rare to have a manganese deficiency.

25

Gathering

From the corners and the edges
and the inner and the outer of the world,
we gather…

-Brian and Wendy Froud, *Trolls*

There is manganese in steel. There is manganese in soda cans. There is manganese in our bones and enabling our metabolism. There is manganese in the pigments of cave paintings! The uses for this humble transition metal are all over the map. But when we look at all the many, many industrial and biological ways manganese is essential, we can see a pattern emerging: Most of the time, what it is doing is *gathering*.

It binds with free radicals in our blood. It de-sulfurs and deoxidizes steel. It bonds with impurities and keeps them from harming whatever process is happening, from glassmaking to photosynthesis. There is much that can be learned from manganese, because we all do tend to go about our days hampered by the little things that make life harder or more complicated, when it is well within our power to simply gather all that up and send it away.

As usual, though, a warning: Above a certain point, this essential nutrient becomes a toxin. It attacks the brain and disrupts the delicate processes there. Knowing what to leave alone is imperative when we go about the crucial act of gathering.

It is time to tidy up. The clutter in your world may be physical or it may be psychological, but in both cases the message of manganese is the same: Gather what you do not need. Release it. Repeat as necessary until you find it easy to wake up every day and live your life. Then stop. Shift gears.

It is good to know what to gather together and keep.

Iron

Element Number:	26
Weight:	55.845u
Normal State in Nature:	solid
Melting Point:	1808° Kelvin
Boiling Point:	2023° Kelvin

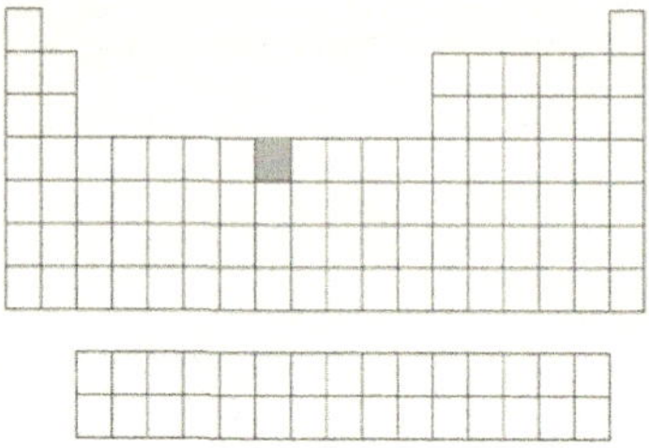

The very core of the Earth is iron. As the most common metal in the world, iron has been used since at least 3500 BC. It is everywhere, and used for almost every aspect of human endeavor. It has an entire historical age named after it. There's never been a war fought over it, it's so common. Most languages have their own word for it. Its Latin name is *ferrum*, which is where we get its abbreviation (Fe) as well as the word *ferrous*, meaning sensitive to a magnetic field.

Long before humans even learned how to smelt it, iron oxides were used as pigments to create red, brown and yellow ochres painted on cave walls. It alloys with many other metals and is the main ingredient in steel. The entire industrial revolution was built on the strength of this one metal.

The only downside to it is that it so easily succombs to oxidation. Archaeologists do not often find iron utensils, because they rust away. Water and air are its nemeses.

Iron is the heaviest element creatable by nucleosynthesis in a functional star; everything larger requires a supernova.

It is a vital nutrient. It is the protein that lives in your blood and carries oxygen to your cells, and it has a dozen other uses in the body, as a building block of enzymes and for nitrogen fixation. It exists in many foods, but it is easy to not get enough. Lack of iron is the most common nutritional deficiency in the world. However, since the body cannot excrete any excess, do not use this as an excuse to go gobble supplements.

Work

You work that you may keep pace with the earth and the soul of the earth.

-Kahlil Gibran, *The Prophet*

If the internet is the brain of our modern world, iron is definitely the bones. The number of buildings and bridges, cars and trucks, pipes and tools that are made of this metal are mind boggling. Even down to the silverware we eat with and the nails that our pictures hang on; there is no job too small for iron. Case in point: the cells of our body cannot absorb oxygen without the help of iron ions, which may be the smallest—but hardly the least important—job it does.

Iron does the heavy lifting in this world, but it doesn't complain. It expects to be handed the tough jobs, the ones requiring both strength and patience. On its own it is a powerhouse with a vast array of skills, but once alloyed with any number of other metals, its uses become endless. It is good to work well with others.

Work is often spoken of disparagingly; people talk as if it's the thing we do when we can't avoid it, as if labor is synonymous with suffering. But iron knows that work is life, and to work well is to live well.

Be aware that working *well* is not synonymous with working *constantly*. Iron can rust, and so can we; the strongest iron structures need to be maintained or they will eventually dissolve. Self care is intrinsic to good work.

If you have drawn this card, you know what to do. Roll up your sleeves and get to work! Dig that garden! Paint that wall! Change that oil! Chop that wood!

Or, as it may be, take those meds. Run that bath. Fight that illness. Work means different things on different days, and if your only contribution today is to fight off rust, that is worthy work.

Cobalt

Element Number:	27
Weight:	58.933195u
Normal State in Nature:	solid
Melting Point:	1768° Kelvin
Boiling Point:	3143° Kelvin

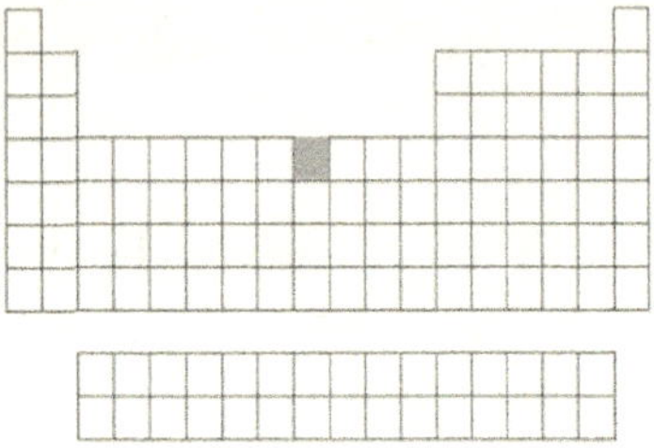

Cobalt was the first of what is considered the modern discoveries of elements. Though it had been used for centuries to color glass and ceramic glazes, anyone trying to smelt the ores it comes in was doomed to get nothing but a pile of dust and a lungful of arsenic. The ores, therefore, were for a long time called by the same name as destructive mine spirits, or *kobolds*.

In 1735, Brandt changed methods and succeeded in proving that it *was* a metal, just not one that behaved as expected. Once isolated, it kept the name, with a new spelling. It is toxic, but useful. It makes for hard, strong alloys, heat-resistant and hypoallergenic. In nature it is stable, but it has 28 radioisotopes. They have been used for imaging and cancer treatment, among other things; but several of its uses are now obsolete.

As a pigment, it is still prized. Most dark blues are cobalt based; it has been in use since the Bronze Age. In glass it is quite beautiful.

It is a component of rechargeable lithium-ion batteries for automobiles and a hardener for steel. Its alloys make excellent high-speed drill bits and machine parts.

Cobalt is the central molecule in vitamin B12, which is necessary for the metabolic function in all animals. It keeps the nervous system running properly and assists in DNA synthesis. Keep in mind that the metal is not equal to the nutrient, of course. Cobalt alone is still toxic, causing contact dermatitis when touched and cardiomyopathy when ingested. Still, it is not as dangerous as its reputation makes it out to be.

27

Forbearance

I can't get too offended when somebody parodies me.

-Weird Al Yankovic

Cobalt has a fair number of radioactive isotopes, and for some people, that is all they want to know. But none of these isotopes occur naturally; the one stable isotope is the one found in nature. Not easy to isolate, mind you, but present. Stable cobalt is a ferromagnetic transition metal. It lends its hardness to high-quality machine tooling. It makes a gorgeous blue pigment, used since ancient times--some antique glass is tinted with it. But there are people who confuse the nuclear fallout with the stable metal, who claim blue paint is radioactive, who avoid cobalt drill bits and predict dire consequences for having it around. To all of this, cobalt just shrugs.

Like the fair folk he was named for—kobolds—this metal doesn't really give a darn about human drama. He just does his thing, and cheerfully ignores any judgement hurled his way. It is easy to show forbearance when the world treats us fairly, but it takes cobalt to have a chill attitude while people are being willfully ignorant, hostile or rude. He doesn't argue, or defend himself. He simply shows forbearance.

It will come as no surprise to you that you are living in a world with a lot of stupid people. This can be frustrating and discouraging, of course, but cobalt is asking you to set that aside for now. Let it go. Ignore idiocy. Forgive them their ignorance, smile to yourself, and keep doing your thing regardless. Keep pouring your color, your strength, your gnomic charm into the world, and give not two shits about the perception of others. Forgive them their lack of perspicacity, and move on.

Nickel

Element Number:	28
Weight:	58.6934u
Normal State in Nature:	solid
Melting Point:	1726° Kelvin
Boiling Point:	3005° Kelvin

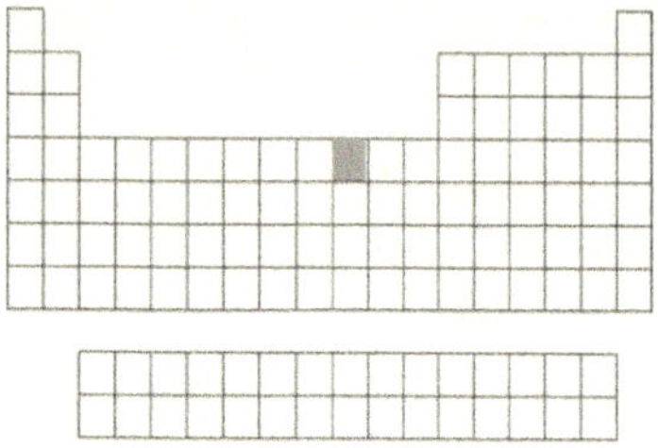

Alongside iron, nickel is a key component of the molten interior of planet Earth. And along with cobalt, it is named for a mythical mine spirit, known for mischief and sabotage. It was easy to mistake the ore for silver or copper, but it did not behave like either, so clearly goblins were involved.

It has been unintentionally used by humans since at least 3500 BCE. Without ever isolating or recognizing the metal element, it nevertheless can be found in ancient bronze relics. Its official discovery, however, was in 1751, by Fredrik Cronstedt.

At room temperature, it's magnetic, but heated to about 355 degrees Celsius it loses that quality. In the 1800s, a nickel coin could have a value of one, two or five cents. In all but minuscule amounts, it's poisonous to humans.

Nickel is used in hundreds of products, mainly stainless steel but also all sorts of other things, from batteries to electroplating. Only a tiny amount is now included in the coin that bears its name, because the metal has become too highly valued. In essence, if it was still solid nickel, a coin worth five cents would cost a great deal more than five cents to make.

Nickel exists in non-dangerous amounts in most settings. Your plated faucets give you a tiny dose, but your kidneys can flush it out. One of the nastiest sources of harmful nickel particulates is cigarette smoke. Over 72 million people are estimated to be allergic to it. In 2008, it was named the Allergen of the Year.

28

Helpfulness

"Where the well-being of the persons to whom one is attached is concerned, one cannot say that one is not responsible," said Sophy simply. "One should make a push to be of service."

-Georgette Heyer, *The Grand Sophy*

Nickel was named for a stubborn spirit of German mining lore because it is so difficult to isolate. Despite the fact that it is a major component of the Earth's molten core, nickel is not at all easy to locate or refine. Doesn't sound helpful, does it? But once it *has* been purified, nickel begins to shine.

It is essential in stainless steel. It is used to plate all sorts of tools, plumbing fixtures and jewelry to make them wear-resistant. It *was* used in coinage, but has grown too expensive for that. It is used in rechargeable batteries. Like helpfulness, it turns up all over the place, keeping rust away, preventing tarnish, being all magnetic and glossy and useful. One of the key reasons it was found to be so hard to smelt was that its ore was often mistaken for silver ore. But nickel does not act like silver, so attempts to work with it were frustrating. As soon as nickel was better understood, the problems stopped. Helpfulness is easy to mistake for other things too. We can all be so cynical. But to recognize helpfulness is to appreciate it and welcome it. Our steel will be stronger, and our spoons will sparkle. We don't have to do it alone. Nickel is here to help.

Yes, nickel is here to help! Take quick stock of your world right now. Is there a weight you are trying to carry alone? Is there someone trying to share the load, but you have not let them? Nickel wants you to know that accepting the assistance is the right thing to do. If that is not your situation (and I know you're being honest with yourself), perhaps it's YOU who is the helpful one. Excellent! Offer your strength, magnetism, and other wonderful qualities to those in need. There has never been a better time to be helpful.

Copper

Element Number:	29
Weight:	63.546u
Normal State in Nature:	solid
Melting Point:	1356.6° Kelvin
Boiling Point:	2840° Kelvin

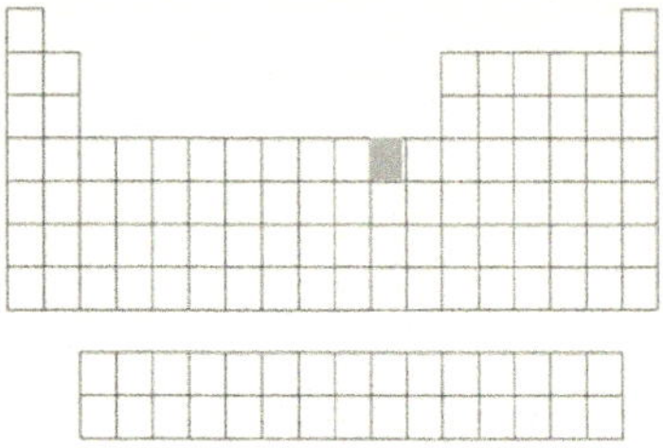

Copper has been used by humans for at least 12,000 years. It was mined by the Romans on the island of Cyprus, and its name is derived from that. It is one of the few elements that can be found in the earth as a native metal. It is antifungal, antimicrobial, probably antiviral, and can be alloyed with virtually any other non-ferrous metal. Bronze is an alloy of copper and tin; bronze is so important, it has an entire age named after it. Brass, a much more recent invention, is copper and zinc. It has an orchestral section named after it.

Ancient artifacts from burial mounds include copper jewelry. Various civilizations throughout history have used it as a currency. Many others used it to make surgical instruments, and to purify drinking water.

As a roof coating, copper will last for centuries. Copper bathroom fixtures are used in hospitals to prevent disease transmission. It is used as wire (because it is electrically conductive), in plumbing, and even to paint the hulls of boats. When exposed to weather, it forms a chalky green patina on its surface, but even that is frequently prized.

A small amount has a critical role in the human body. It aids in electron and oxygen transportation. It can't be absorbed through the skin; trace amounts need to be included in the diet, because it is absorbed through the gut and redistributed by the liver. It facilitates iron uptake, so anemia might be a sign of a deficiency.

Copper should be the official metal of the human race.

29

Humility

My wish is to stay always like this, living quietly in a corner of nature.

- Claude Monet

Consider copper. As a metal, it is impressive. It is one of the earliest to be discovered and worked by humans; it is a component of bronze and brass, both very important alloys. Some of the oldest artifacts that exist are copper jewelry, worn by ancient royalty. It is an excellent conductor of electricity, and has a thousand uses in our modern world. Also, both clean and patina'd, it is beautiful. And yet, with all this pedigree and power, copper is not jealous of its dignity. It has been the humble penny, the simple basin, the helpful downspout. Copper does not make itself difficult to find or work. Rather, it welcomes change, heat, and the inclusion of other metals without a fuss.

Humility is often described as a belief that others are better than us, or that we are insignificant. This is a sad misconception. In reality, humility is to be content in the knowledge that we do have worth…but so does everyone. And some things are smaller than us, but many things are bigger. And that we each have strengths, but we also have weaknesses. We can fail, but when we have humility, we do not fear failure. Like copper, we do what we can, accept help when it's offered, and never get hung up on being the highest value of coin.

With copper as your guide, it is time to practice humility. Turn down your self-importance. Stop thinking of your accomplishments, your failures, even your wounds. Let other things and people be as important as your own stuff. Look outward, not inward. Bow your head at the sheer enormity of the universe. You are a small but necessary part of creation. So, accept help if you need it. Never stop welcoming change. Be grateful. And be humble.

Zinc

Element Number:	30
Weight:	65.38u
Normal State in Nature:	solid
Melting Point:	692.73° Kelvin
Boiling Point:	1180° Kelvin

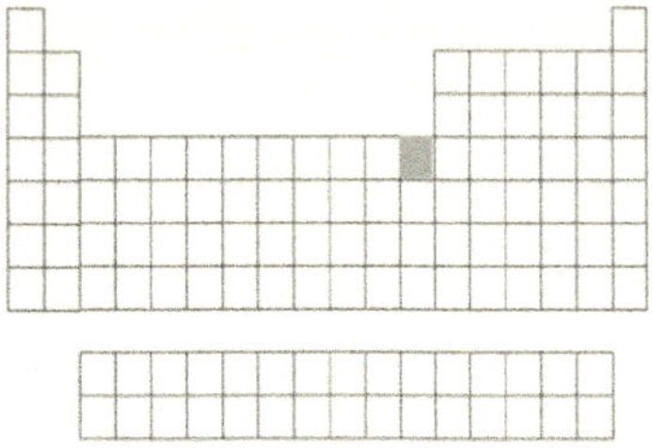

One of the very oldest metals known, zinc was used by just about every culture in just about every way. Its alloy with copper is brass, which has existed for thousands of years; and metal artifacts that are mostly zinc have been found which are technically prehistoric in origin.

It has had many different names throughout the world and the centuries. As far as anyone can tell, Paracelcus was the first to refer to it as zinc (zincum, actually), in his 16th century book of alchemy. It might be derived from German *zinke*, meaning spiky; or possibly mean "similar to tin" since the German for tin is *zinn*. It could even be based on the Persian word for stone. Only Paracelcus knows.

Wherever its name comes from, zinc has a mind-boggling array of uses.

It is a vital nutrient for humans, second only to iron in the list of metals in the human body. We need to eat it to live. And not just us animals; zinc is crucial for plants, micro-organisms, even coral. It has dozens of biological jobs to do, and a deficiency can kill.

But that's just the start. Zinc shows up as a rust inhibitor on metal roofs, as a UV shield in sun block, on galvanized steel hardware; in batteries, deodorant, dandruff shampoo, and glow-in-the-dark paint. It's the base metal in pennies. Zinc anode rods protect your water heater by taking on the oxygen that would degrade the steel shell, keeping it occupied. Bricks of zinc are attached to steel structures for the same reason.

If you mix its powder with sulfur, you can make it go boom. But don't.

Talisman

A lovely thing of blue and gold and red that shone like a star in her hand..."If you ever need me, as often as you need me, wave this three times in the air, and I will come to you, unless death prevents me."

-Mercedes Lackey, *Firebird*

Impure zinc has been around for a very long time. In every ancient civilized culture, it seems, zinc made itself useful in alloys, such as brass vases in India, and compounds, including zinc carbonate used medicinally in Rome. That was very wise of them, as zinc is a vital micronutrient for humans, used by the body for normal daily processes as well as to stave off illness and depression. Plus it blocks UV light, so no sunscreen is complete without it. Like a talisman, which, when you carry it, prevents bad things from happening, zinc is the protector of body and soul.

When it steps away from organisms and into its more technical uses, zinc does not stop being a talisman. Because it attracts any oxygen in the area, its presence keeps other metals safer from corrosion. Iron bridges, steel water heaters, ships' rudders, and such are kitted out with a piece of zinc somewhere—a talisman—that will dissolve as it protects, like a single-use amulet in a fantasy story. Of course, like any practical magic, zinc does have its downside; when there's too much in the water or soil, it still does its thing—hoard oxygen—which can effectively suffocate any plants or fish trying to live there. Such tragedies are the result of industry's carelessness, and preventable. It is best not to lose track of our talismans.

What you need right now is a talisman. Some small physical thing, some item from the back of a drawer or a thrift store shelf, a pretty stone you found on a walk, a piece of jewelry extracted from the tangled mess in a box, you know the one. Wear it or carry it; keep it handy for a while. Imbue it with the power to protect you from corrosion of the soul. If you eventually lose it, or forget it, good! It has done its job. Meanwhile, carry your talisman and wear your sunscreen.

Gallium

Element Number:	31
Weight:	69.723u
Normal State in Nature:	liquid
Melting Point:	302.92° Kelvin
Boiling Point:	2478° Kelvin

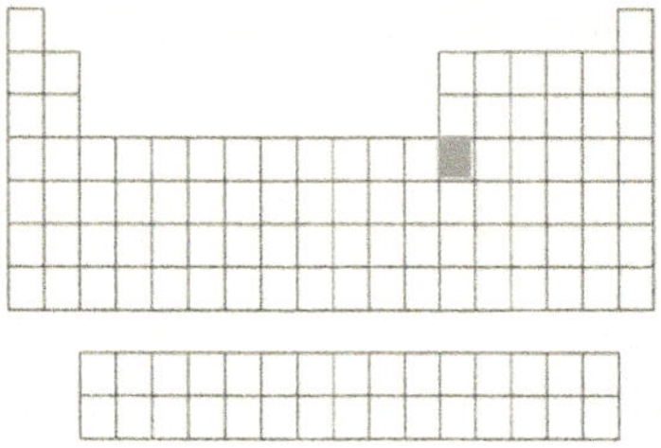

Gallium was discovered by de Boisbaudran in 1875, using spectroscopy. Mendeleev had predicted four years earlier that it would be found that way. He had also predicted everything about gallium's physical characteristics, from its unusually low melting point, to its density, to the PH of its oxides. So while it was certainly discovered and named by de Boisbaudran, it was also a triumph for Mendeleev.

De Boisbaudran named it *gallia*, after his native Gaul, another word for France.

Since gallium is never found outside of its various mineral compounds, its availability is determined by how much zinc and bauxite (aluminum ore) are mined. It is extracted from those ores as a sort of by-product of the more important metals being isolated.

Historically, it was mainly used to lower the melting point of metal alloys.

These days, 98% of all gallium produced is used in semiconductors. It is a technology-critical element; almost all electronics require it. Smartphones, laser diodes, computer chips...You name it, it needs gallium. Blue LEDs are gallium-based.

A piece of gallium in your hand will melt. It is very shiny and slippery, and is used to make mirrors, and ski wax. Its behavior is similar to mercury, and it is used in thermometers. But it differs from mercury in at least one very important way: It is non-toxic. While holding it in your bare hand will stain your skin brown, it will not harm you. Any of the metal that you may accidentally absorb will be safely excreted.

Don't eat it, though.

31

Innocence

In a sense, Judy got away with murder.
In a sense, Elvis lived.
In a sense, innocence.

-Kinky Friedman, *Elvis, Jesus and Coca-Cola*

There aren't a lot of elements that are liquid at normal pressure and temperatures, and they tend to be dangerous. But where bromine burns, and cesium explodes, and mercury leads to madness and death, gallium is just…fun. Once teased out of the minerals containing it, gallium is a soft, supple metal that melts at skin temperature. It is used in thermometers since it is both safer and more effective than mercury, and in alloys, to lower the melting point of the whole. But it is also perfect for science experiments with children, to cast a tiny figure in a mold and liquify it again with a touch; to see little faces light up when they realize they're holding molten metal in their hands.

Like innocence itself, gallium is both nontoxic and uncomplicated. It is not an essential nutrient, but it *is* technology-critical. Innocence illuminates, like the blue and purple LEDs that use it, but does not flood with light. It is a candle in the dark, not a blowtorch. Gallium can be lost over time to entropy, or it can harden incorrectly and become brittle and shatter. Innocence is also subject to these dangers, as well we all know. But the good news is, it can be gathered up again; and once it is warm, it flows back together, and becomes whole.

Can you open your eyes to the universe in childlike wonder? Can you jump in a mud puddle, dance in the rain, build a blanket fort and eat dinner backwards? Gallium is inviting you to do all these and more. Look at life with wide-eyed delight. Even—especially—if this feels silly. Do you think you've lost that capacity? Gather the shards of your innocence and hold them awhile. Can you not find them? Do you think that innocence is lost for good? Perhaps you just don't recognize it because your soul is using it as a light in the dark…

Germanium

Element Number:	32
Weight:	72.64u
Normal State in Nature:	solid
Melting Point:	1211.5° Kelvin
Boiling Point:	3107° Kelvin

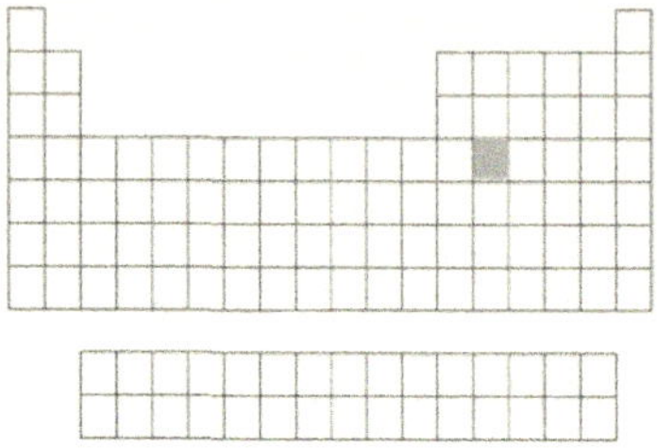

Germanium is one of those elements that was found after Mendeleev predicted it. In this case it took 17 years for the discovery to follow the prophecy. Winkler was the one who finally found it, in 1886, in the mineral argyrodite. He named it for Germany, his country. It is found in conjunction with other metals, especially aluminum and zinc, but also sometimes with silver and lead.

It was thought to be just a poorly conductive metal, until the invention of semiconductors suddenly threw its real skills into the limelight. It was in early transistors in the 1940s, and used in virtually every transistor radio ever made. The world's first microchip was printed on germanium. Most of the technological advances of the mid-20th-century required it, and it is still considered technology critical.

Many of its early uses eventually got usurped by silicon, because, while silicon needs to be much purer than germanium in order to be utilized in electronics, there is so much silicon available that it is simply more sustainable to go that route. Germanium is still important, though. Its main uses today are in semiconductors and fiber optics. It can be found in solar panels, resistors, infrared optics and LEDs. Its radioactive isotope is used in chemotherapy.

It is even possible to synthesize nanotubes out of germanium, so near-future nanotechnology could well rely on it as heavily as current technology does.

It does nothing in your body, but some compounds are hard on the kidneys.

32

Innovation

"Contrary to popular belief, I know exactly what I'm doing."

-Tony Stark, *Iron Man 2*

Germanium is an element of ideas. When, in the mid-20th century, bright minds began inventing transistors and diodes, it was germanium that they used. Even the first microchips were printed on it. Wherever innovation was happening, germanium was in the thick of it. And this is still true; germanium is technology critical. It is being used today in semiconductors, solar panels, LEDs, fiber optics and nanotubes. Whatever is cutting edge, germanium is there.

Innovation can take many forms, but what matters across the board is enthusiasm for discovery. Not only is germanium exemplary of this, but it also has the crucial distinction of being a normal metallic substance to visible light, but transparent to infrared. This makes it incredibly useful in optics and thermal imaging; just as a simple idea may be opaque to an ordinary mind, but provide a spectrum of information to anyone willing to look with the eyes of innovation.

If germanium is in your spread, this is the time for ideas. More: It is the time to pursue those ideas until they become real. Tinker around, manifest something, channel your inner inventor. Build a better mousetrap. Share the creations and connections you've made with others, because you may well have the solution to someone else's roadblock. And innovation, when shared and multiplied, can change the world.

Maybe even save it.

Arsenic

Element Number:	33
Weight:	74.9216u
Normal State in Nature:	solid
Melting Point:	1090° Kelvin
Boiling Point:	876° Kelvin

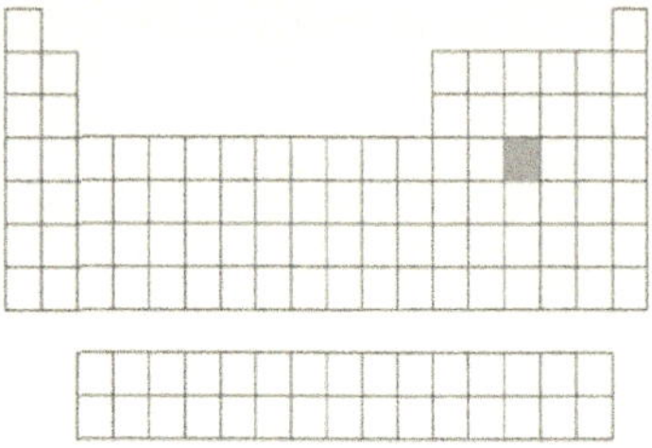

Arsenic has been with humans since antiquity. It name can be traced all the way back to the Aramaic language, and was at first the name for one of its minerals, not the isolated metalloid itself. It was first isolated by Jabir ibn Hayyan around 815 C.E.

It is one of the most poisonous elements known to humans, in any form. Its main use over the centuries has been as a murder weapon. Poisoning your enemies (or superfluous family members) with arsenic is an old tradition, especially amongst the aristocracy; so much so that arsenic has been called the Poison of Kings.

Some of its uses are benign, such as the hardening of bronze or as a treatment for parasites. This has, many times, lead people into the error of using it in things like face powder and wallpaper pigment, resulting in an uncountable number of accidental fatalities.

The boiling point of arsenic is lower than the melting point in room air, so it only melts when it is under pressure. Airborne arsenic has a garlic odor, which is sometimes noted in groundwater, where it's common. This makes thousands of people sick every year. Until the Marsh Test was invented, it was untraceable.

What arsenic does best is kill you. It has been banned in most places for most uses because it is so deadly. Of course, there will always be ingenius people who think they can outsmart toxicity. The brilliant green pigment once used in wallpaper has been reinvented as a pesticide. As recently as 2003, arsenic was the wood preservative of choice, and even now it is used in the U.S. as a poultry feed additive, to inprove the birds' weight gain.

33

Misdirection

Solutions nearly always come from the direction you least expect, which means there's no point trying to look in that direction because it won't be coming from there.
-Douglas Adams, *The Salmon of Doubt*

Why is she carrying a chicken? So you won't ask about the bag of skulls. Okay, why is she carrying a bag of skulls?

So you won't ask about the chicken.

Arsenic has a long and complicated history with humanity. We know that it is poison, but we keep on finding handy uses for it that, whoops, turn out to have been bad ideas. Why did this racehorse die? Or the girl in the tarlatane gown? Why does the damp air in the green drawing room cause terrible illness? Heavens, we have no way of knowing…

When one is the master of misdirection, one can control what people see.

One reason for this historical blindness is that arsenic has so many compounds with different properties. A few of its uses really are harmless. Since antiquity, it has been used to make bronze harder, for example. The modern age just made it easier to muddy the water. How many of us are really aware of what goes into the objects in our daily lives? The good news is, these days, arsenic is no longer used to treat wood for playground equipment, or as medicine for syphilis. But it is still used in car batteries, and in industrial chicken feed. Surely, neither of these will turn out to have been a mistake, right? Surely we are too sophisticated now to make the mistakes of our ancestors…

Arsenic is a warning that your perspective is being adjusted by sinister forces. Whether the smoke-and-mirrors act is coming from a person, an information stream, or even your own assumptions, it is time to confront the fact that it is toxic. Shake it off. Even if wrenching yourself loose makes you feel confused. Even if it was more relaxing to fall for the lie. Free yourself and see what is really going on. Before it's too late.

Selenium

Element Number:	34
Weight:	78.96u
Normal State in Nature:	solid
Melting Point:	494° Kelvin
Boiling Point:	958° Kelvin

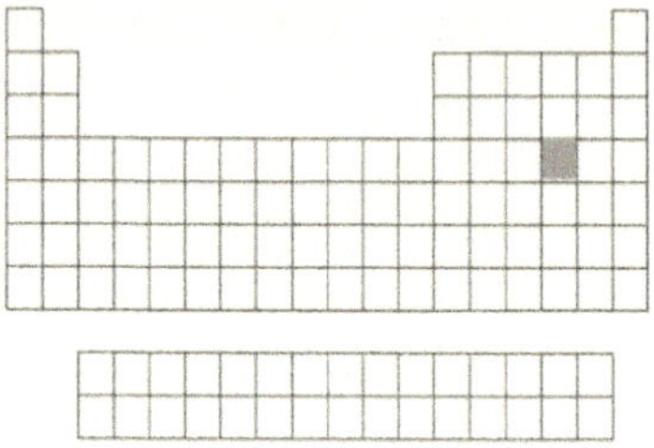

Selenium was discovered by Berzelius and Gahn in 1817, who noticed an unexpected red precipitate occurring in their chemistry plant while producing sulfuric acid from pyrite. At first they mistook it for arsenic, and discontinued production; then they were sure it was a tellurium compound. Eventually they realized they had found a new element, closely related to tellurium, and named it after the Moon.

In 1873, selenium was found to have a variance in its electrical resistance depending on ambient light levels. This is such a unique trait, it was put to use in photoreceptive inventions right away. Light meters, rectifiers, all 20th century xerox technology...Even what Alexander Graham Bell considered his greatest invention, the photophone, which transmitted voices on light waves and was way ahead of its time.

The first rooftop solar collector used selenium, and is still an integral part of the cells. It is used to replace the former lead in brass, especially plumbing fittings. A minuscule amount is needed in our thyroid glands. But keep in mind that it is a powerful poison to fish and waterfowl, as well as people. Astragalus (locoweed) is a flag for high selenium in soil.

Garlic breath when you have not eaten any garlic is a warning sign that you may have overdosed on selenium. It is found in dryland areas such as adobe soils. The poison water in bad westerns was often due to selenium (or arsenic) salts. Planting astragalus and woody asters can help leach it from the soil.

34

Balance

"Holy, holy, holiest of sisters, who is both black and white, both dark and light, your coming is the beginning and it is the end."

-Jane Yolen, *Sister Light, Sister Dark*

Many elements are two-edged in that they are essential in some amounts and deadly in others, but none has a narrower margin than selenium. It has been claimed, in fact, that everyone everywhere has a chronic selenium deficiency, except the few who are overdosing. It is very rare to find someone who has achieved the proper equilibrium, and in most cases they luck into it. Balance can also be a matter of instinct rather than exhaustive thought or effort.

How can you tell your personal selenium state? Test the soil. There is a strong correlation between a person's selenium level and that of their location. So start with the earth, and an awareness of the earth. To have a feel for ourselves within our space is a powerful thing. It makes it much more difficult for circumstances to knock us off balance.

High-selenium soils are a favorite haunt of a plant that ranchers call locoweed, because eating it makes cows go mad. Selenium overdose will literally unbalance a brain. Knowing the signs of the things that may rob us of balance is another point of power. Knowledge, observation and awareness make balance that much easier to achieve.

When you draw selenium, it means it's time to pay attention to your sense of balance. What extremes are you hovering between? What burdens are you carrying that may be knocking you sideways? Breathe deep and focus first on the earth that supports you, then on your own center. The goal is not a static state, but a dance. Balance is to be found in movement, in feelings and in trusting your intuition.

Watch out for the crazy flowers.

Bromine

Element Number:	35
Weight:	79.904u
Normal State in Nature:	liquid
Melting Point:	265.95° Kelvin
Boiling Point:	331.85° Kelvin

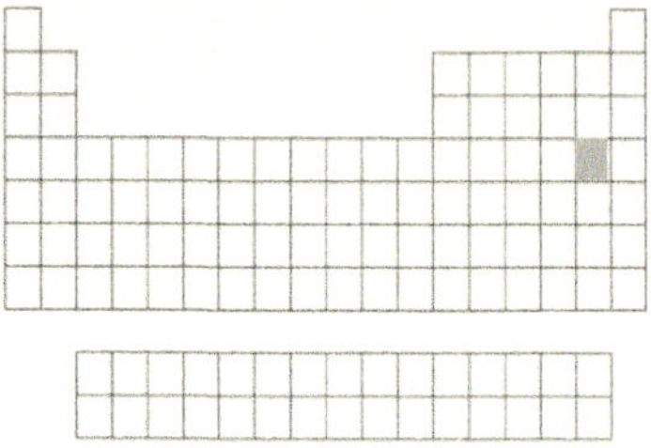

Bromine was discovered simultaneously by Löwig and Balard in the mid 1820s. Löwig delayed publishing his results, so by default Balard got to name it. Since he had isolated it from the ash of salt marsh seaweed, he at first called it *muride* after the Latin word for brine. His contemporaries talked him into renaming it bromine, meaning "stench."

Bromine is the third of the halogens, directly below chlorine, right above iodine and midway between them in chemical properties as well. At standard temerature and pressure it is a liquid, reddish brown in color and highly reactive. It forms bonds readily to make various salts and acids, and even bonds with itself when left alone.

Early uses for it included sedatives, daguerreotypy and, during the first world war, as a weaponized gas.

For a while during the height of its popularity, bromine was used for a lot of things. As well as being a sleep aid, it was a gasoline additive, an ingredient in soft drinks, and an insecticide. In the early 1900s you could order a "Bromo" at the drug store for a head-ache or indigestion. Its use was sharply curtailed due to our discovery of its tendency to destroy the ozone layer. The Montreal Protocol, starting in 1989, phased it out.

We therefore no longer tent houses and fill them with bromine gas to kill termites, but bromine is still quite important as a fire retardant, sometimes on children's pajamas. It has other uses in analog photography, batteries, and as an antimicrobial additive for hot tubs and spas.

35

Sword

I do not love the bright sword for its sharpness, nor the arrow for its swiftness, nor the warrior for his glory. I love only that which they defend...

-Faramir, *Lord of the Rings*

Bromine is not a straightforward element. As a halogen, it is toxic, caustic and very bad to breathe. It can do severe damage in the wrong hands. Like a sword. But also like a sword, it can be a powerful tool for the protection of the weak or innocent, and that is a fine place to focus. Brominated salts are used as a disinfectant in hot tubs, protecting people from harmful bacteria that would otherwise thrive there. This task belongs to bromine both because it can handle the high temperatures and because, unlike chlorine, it will not harm a delicate lawn when the tub is emptied for cleaning.

But a far more prevalent job for bromine is as a flame retardant. Circuit boards, polystyrenes, even some epoxy resins are treated with various forms of it; in fact, nearly every situation where a fire must not be allowed to burn. Kids' pajamas in particular are treated with a bromine compound—by law, in some places—to protect children in the event of a fire. Until a few years ago, bromine was used to fumigate for pest control, but that proved to be stabbing holes in the ozone layer, and the practice was laid to rest. There are better ways to protect our world, and, like bromine, we must be vigilant against humanity's tendency to--literally or metaphorically--set itself on fire.

First and foremost: Do you need protection? Or do you need to serve as protection? There are many hazards in our world, and if you are feeling vulnerable, bromine is your cue to reach out—to a trusted friend, a reliable authority, or a can of mace. Whichever is called for. Get yourself a sword. Conversely, if you are coming from a place of power and confidence, look around you. Someone needs your protection right now, and good things can happen when you pick up your sword in defense of the defenseless.

Krypton

Element Number:	36
Weight:	83.798u
Normal State in Nature:	gas
Melting Point:	116° Kelvin
Boiling Point:	120.85° Kelvin

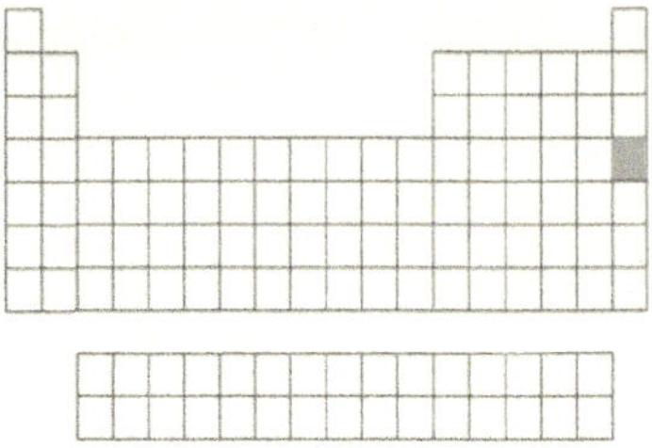

The discoverer of krypton, William Ramsay, also isolated most of the other noble gases, and got the chemistry Nobel prize in 1904 for his work. He named this one for the Greek word for "hidden." Krypton has existed since the universe was formed, and while we know that its concentration in the Earth's atmosphere is 1ppm, it is very hard to measure how much is in the universe as a whole. There is probably a lot, but we use meteoric activity to find it and that can be unreliable. Hidden indeed.

The decay of one isotope is used to measure the age of ancient groundwater. For a while in the 20th century, the International Bureau of Weights and Measures used krypton to define both the meter and, by extension, the angstrom. This definition served until 1983.

Krypton is mostly used to make light, in all sorts of ways. Its multiple, bright emission lines mean that the light it casts appears to be intense white, so it is utilized in high-speed photography flashes; it is added to fluorescent lights to amp up their efficiency, used in electrified tubing-type lights to make an aqua color, and is used instead of a vacuum in expensive incandescent bulbs to help the filament last longer. Lasers, both for entertainment and for research, rely on krypton.

When you breathe it, it acts like a narcotic, and, yes, could conceivably stop your normal breathing. It could make you "faint of heart."

It doesn't affect Superman.

36

Illumination

Until, from the midst of this darkness, a sudden light broke in upon me--a light so brilliant and wonderous, and yet so simple…

-Mary Shelley, *Frankenstein*

Laser light shows, expensive lightbulbs, photography flashes and luminous signs; one cannot deny that krypton has a calling. There is some irony in its name meaning "unknown" or "dark," because all it does is bring the light—white, bright and blazing. Whether it's glowing in its own right as an electric current passes through it, or enveloping a tungsten filament to facilitate a strong shine, krypton is the master of illumination.

The light of knowledge it represents is as important as the literal light it sheds. Knowledge is power, they say, but it is also many other things. It is empathy: The knowledge of the Other. Or it is choice: To know one's options and which is wisest. It can even be joy: to gain knowledge is to learn and to learn is to grow, like a tree in the sunlight. Light, again, is key.

Like the other noble gases, krypton does not react or bond with any other element. Illumination of the mind is also neither dependent on other circumstances, nor does it directly influence them. It merely gives one the power to enact change, or spread awareness. Its role is cause, rather than effect. It belongs at the beginning of things, to light the path forward, to illuminate the hitherto unknown.

Krypton is in your spread to give you the means to see things that were hidden. Let its glow illuminate the work you are doing, the places and people with whom you interact, the problems you are wrestling. Look hard at all of these in the steady, bluish-white brilliance shed by krypton. What were you missing? What's becoming clear? Have yourself a good "A-HA!" moment and move forward with the confidence that knowledge brings. It is much easier to function in the light.

Rubidium

Element Number:	37
Weight:	85.4678u
Normal State in Nature:	solid
Melting Point:	312.63° Kelvin
Boiling Point:	961° Kelvin

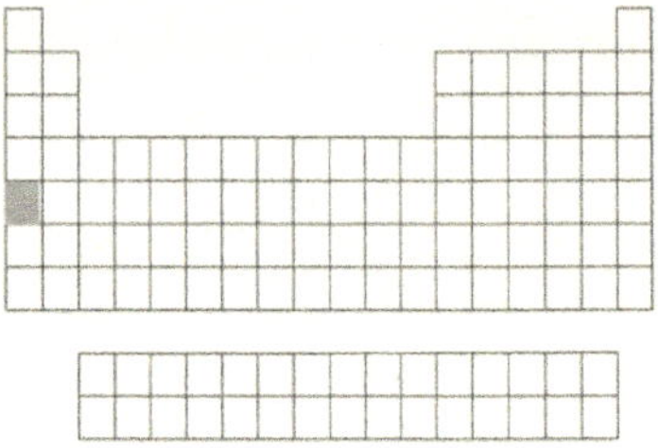

Flame spectroscopy was invented by the scientists Bunsen and Kirchhoff in 1860, and within a year they had used it to discover two alkali metals: first cesium, then rubidium. They named rubidium after its deep red spectral line, and it took a few tries to isolate. Their first effort turned up, not an alkali metal, but a strange blue substance that was probably a rubidium chloride compound. But they did finally manage a pure sample.

Once isolated, rubidium did not seem worth the trouble. It is pyrophoric, too rare to ever be less than expensive, and its uses are sparse. It has no biological role, although animal cells exposed to it will treat it the same way as potassium, which it does resemble chemically. The few industrial applications it has require only a tiny amount.

It has both a stable and a radioactive naturally occurring isotope; but since the half-life of the latter is three times the age of the universe, it can all be treated as effectively stable.

It makes fireworks purple. It's used as a marker for brain tumors. Rubidium vaporizes easily, and this vapor vibrates at a reliable rate; so it is used as the resonant element in atomic clocks, as well as in physics labs for laser cooling and bose-einstein condensation.

Really, its biggest application is in physics research. It has a dozen uses in that field alone. But not much anywhere else. Yet.

Dexterity

World-famous artistes performing death-defying feats of dexterity and skill before your very eyes!

-Bert, *Mary Poppins*

Rubidium has a whole lot of applications, very few of which are lucid to the layman. It can be found in laser cooling and Bose-Einstein condensation, in polarization of helium and in vapor turbines. Not many of us know what any of these things are, but we can at least grasp the fact that rubidium is a well-known prestidigitator in science circles. Two of its jobs do stand out as straightforward: It is used in purple fireworks, and in very accurate clocks. Such enormous range calls for impressive dexterity.

Dexterity does not just mean an ability to perform acrobatics, although that certainly counts. It is a mental nimbleness that one only gets by being well grounded. Rubidium, whatever form it takes, knows itself; and it can do all of its jobs with grace. Its various uses are due to its long list of traits. Because it is an alkali metal, its salts make pretty explosions. Because it vaporizes when warmed up, it can be used to suspend movement in other atoms. Because, as a vapor, it vibrates at a very specific frequency, it can be used to measure time. Because of its close relationship with potassium, it can be used as a marker for biologists to study potassium behavior in the body. Like a character actor or a really good substitute teacher, rubidium does many things, and does them well. As can we, when we know our own strengths and abilities.

When you draw rubidium, it is a sign that your gifts, skills and passions should all be getting some air. Practice shifting between them until your transitions are smooth and easy. Remember to focus on each activity as you do it, to give it your full attention as long as it lasts. You may be doing some delicate, time-sensitive task one minute, then exploding into purple fire the next. As long as you do both with skill, you are embodying rubidium.

Strontium

Element Number:	38
Weight:	87.62u
Normal State in Nature:	solid
Melting Point:	1042° Kelvin
Boiling Point:	1655° Kelvin

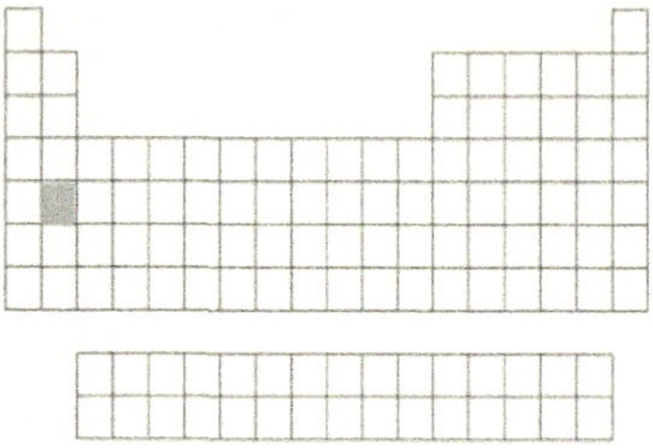

Starting in 1790, a string of physicians, geologists and other interested parties took note of the fact that there was something strange about a certain mineral coming out of a mine near Strontian, Scotland. First Crawford and Cruickshank, then Sulzer and Blumenbach, Then Thomas Charles Hope, all wrote about this "earth" and how it was possibly a new element. Finally in 1808, Davy proved them all right, and named it strontium after the town where it came from.

Strontium hydroxide found its calling in the production of beet sugar, a process invented in 1849 and used for something like a hundred years.

It was embedded in the glass screens of old CRT televisions, to block the transmission of X-rays.

Strontium is so similar to calcium, bodies uptake it the same way and store it in the bones. This makes radioactive strontium useful in treatment of bone cancer, because it will concentrate the radiation where the degeneration is worst. That radioactive isotope does not occur naturally, but is a product of nuclear fission.

Stable strontium's uses include making red fireworks, toothpaste for sensitive teeth, and glow-in-the-dark toys.

Its quirk of being mistaken for calcium and getting absorbed by bones, as well as the ratio of its several stable isotopes acting like a fingerprint for location, make it useful for archaeologists learning where ancient remains are from, and where they've been.

38

Confusion

"And frogs will fly on mayday morn,
while fishes sing aloud at dawn."

-Brian Jacques, *Mariel of Redwall*

When strontium is ingested, the body confuses it with calcium and sends it to the bones. In most cases this is not proven to be particularly good or bad for the body; but every once in a while, when it happens with strontium's one radioactive isotope, the outcome is pure calamity. What many people think they know about strontium is that it is radioactive, but once again, this is a result of confusion. All naturally occurring strontium is perfectly stable. It takes human ingenuity to turn it into fallout.

Confusion is a familiar feeling to all of us, when we are given incomplete information—or the wrong information—or too much information. We can all relate to that sense of displacement. It is an inevitable part of life, and it makes us feel, like strontium, both reactive and easily oxidized—that is, when confused, we explode, or get covered in a thick layer of ennui. In short, we do not function at all well. It is not easy to overcome the paralysis confusion can cause.

One solution is, like strontium, to lean into it. So we are confused. So be it. One does not need to feel informed and in control at all times. Some of us cannot be confused without getting angry, but that is as foolish a reaction as exploding. It is better to laugh, shrug, and feel out our options. Eventually, we'll get it; the fog will clear; we can move forward. Until then, this is simply where we are.

Strontium is here to help you come to grips with your confusion. The world is not making sense, but that's okay. You may be floundering around in a sea of conjecture and uncertainty, and that's okay. You may just have no idea what's going on or where to go from here, and again, that is just okay. Just keep doing what you do, stay true to yourself, and hang in there. The stable isotopes will win out in the end. The blindfold will be removed. This too shall pass.

Yttrium

Element Number:	39
Weight:	88.90585u
Normal State in Nature:	solid
Melting Point:	1795° Kelvin
Boiling Point:	3611° Kelvin

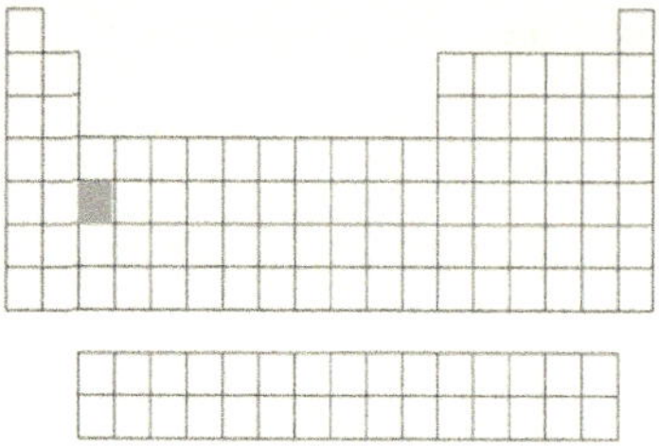

It was 1787 when an amateur chemist named Arrhenius found a strange, dark mineral near Ytterby, Sweden, and sent it to various scientist friends for analysis. The mineral was, over several years, found to contain many previously undiscovered elements, all of which were eventually named after Ytterby; the first of these was Yttrium. Gadolin reduced it to its oxide in 1789, and Wöhler isolated the metal in 1828.

It is sometimes considered a rare earth element, since it occurs in the same ores as the lanthanides and, like them, is never found as a free element. Other times it is referred to as a transition metal, because it exists in the D block and is a metal.

As with most things related to chemistry, its definition is fluid, and depends on circumstances.

Yttrium provided the red in the screens of CRT televisions. Current LED technology also requires it. It is used to make synthetic garnets, such as YIG (yttrium iron garnet), an excellent microwave filter, and YAG (yttrium aluminum garnet) which is crucial for certain lasers. Yttrium is used in anti-cancer drugs and radiation treatments, and it makes awesome superconductors. Small amounts of it act as a material enhancer, making alloys better across the board.

Yttrium does nothing in your body, but it is present in most vegetables, of which cabbage is the champion. And, to top it all, it is a major player in *YInMn*, the first blue pigment discovered in 200 years.

39

Magic

Magic is the bloodstream of the universe. Forget all you know, or think you know. All that you require is your intuition.

-The High Aldwin, *Willow*

A small amount of yttrium added to any one of a long list of other elements will make the strong ones stronger, the tough ones tougher, the bendy ones more bendy. Somehow, its inclusion simply allows the other elements to be more themselves. And what is magic but an enhancement of our own identities, talents and passions? What is magic if not the natural world made bold?

Yttrium is not simply a spell of enhancement. It is active in its own right. Yttrium aluminum garnet lasers are what scientists bounce off the Moon, as and when they need to. Yttrium barium copper oxide is a superconductor, which means that when cold enough, it begins to have very strange and, yes, magical properties. Radioactive yttrium isotopes are used to treat cancer, so even healing magic is in its skillset. And, while CRT televisions have been replaced by LED screens, both require yttrium's presence. Technology changes, but magic endures.

Is yttrium more magical than any other element? Actually, no. Every single one has got properties that no one else has; all contribute something to this mad, beautiful universe to make it more mad and more beautiful. But yttrium invites us to step back and look at the marvels surrounding us every moment of every day, and say, Gosh. Magic *is* real.

With yttrium in your spread, you are cordially invited to experience the magic inherent in the universe, and in yourself. Conduct a lightning storm. Cast a protection spell. Cook a mystical meal. Observe a bumblebee. We cannot spend all of our time marveling at the awesome interactions between thought and matter, but today, please do just that. Bounce your soul off the Moon.

Zirconium

Element Number:	40
Weight:	91.224u
Normal State in Nature:	solid
Melting Point:	2128° Kelvin
Boiling Point:	4682° Kelvin

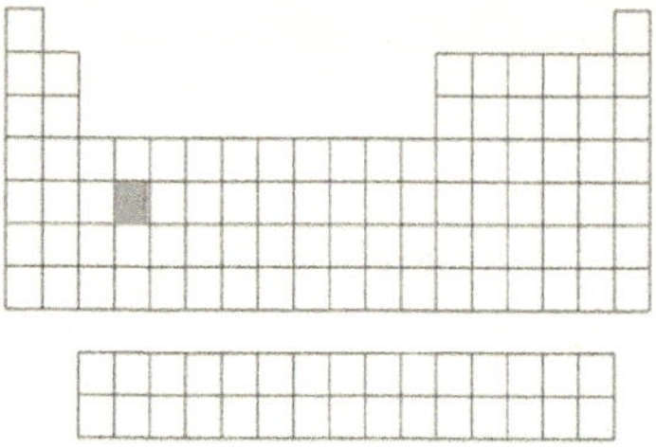

The mineral zircon, familiar to humans for thousands of years, was found by Klaproth to contain a new element in 1789. Davy, for once, failed to isolate it; a mostly pure sample was achieved by Berzelius in 1824. It was not available in industrial amounts until the early mid 20th century, when processes were developed to separate it from its minerals. Even then, the minerals have more uses than the isolated element.

Said naturally-occurring minerals, aside from zircon, include jargoon, hyacinth, ligure and jacinth. They can be quite beautiful and have a long history of being used in jewelry. Synthetic cubic zirconium (ZiO_2) was first successfully produced by Soviet scientists in the mid-1970s, using a method with the delightfully macabre name of *skull crucible* process. This method is still used today.

The multi-hued crystals used for jewelry are almost all synthesized this way. The mother mineral, zircon, is used in ceramics because of its hardness; ceramic knives are made from it, as are angle grinder abrasive discs.

Zirconium itself turned out to be a strong, ductile, and corrosion-resistant metal, excellent for imparting its traits to alloys used in surgical equipment and medical implants, as well as high-end watch cases. Basically, any place you need a really hard metal, look for a zirconium alloy.

It also binds urea, making it essential in dialysis. It is inert in your body, so is a go-to for dental implants and joint replacements.

Delight

"By my faith," quoth Robin to himself, "I do verily believe that this is the merriest feast, the merriest wight, the merriest place and the merriest sight in all merry England. Methought there was another here, but it must have been this holy man talking to himself."

-Howard Pyle, *Robin Hood*

Zirconium is so tough, so strong and abrasive and enduring, that one might think that was all there is to it. It's a transition metal, after all; those guys are all mean, lean and structural. But when it pairs up with oxygen, zirconium shows its hidden colors. Cubic zirconia is a lovely gem, which comes in all hues, costs very little, refracts light beautifully, and is just an absolute delight.

Zirconium can enjoy such a wide variety of activities. Angle grinding? Sure! Shielding nuclear reactors? Absolutely! A gemstone necklace? Let's go dancing! It has applications in space exploration, in surgical tools and in laboratory crucibles, but is also perfectly at home in earrings and tiaras. It takes a certain capacity for joy to thrive in such disparate places, and zirconium embodies that. Be delighted, he says. Be where you are, and enjoy it. Whether your circumstances require toughness or prettiness, be that thing. Why not? Why not take delight in your glorious skin?

Zirconium wants you to smile. Breathe deep, and look around. Where are you? What is going on around you? What would you like to be doing? You have a lot of options; people are more complex than we allow ourselves to be. Ask yourself this: What, in this place and time, would bring me delight? Which of my many selves could thrive here? Shall I be strong and opaque and happy, or translucent and scintillating and happy? Embrace your capacity for delight. Be zirconium.

Niobium

Element Number:	41
Weight:	92.90638u
Normal State in Nature:	solid
Melting Point:	2742° Kelvin
Boiling Point:	5015° Kelvin

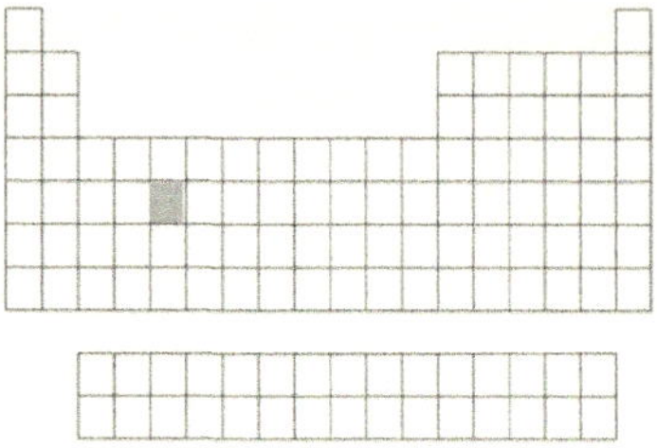

Niobium and Tantalum are so closely linked, so similar in chemical properties and so intermixed in their source minerals, that scientists argued for decades about whether they were a single new element, or two, or even four. The mineral containing them was found in Connecticut in 1734, but the true nature of the metals in question was not established until 1864, by a handful of scientists, including Blomstrand, Deville and Troost. De Marignac in 1866 confirmed all their findings, and the waveform, so to speak, was finally collapsed. Since the name of tantalum had been established years before, and the conjectural metals all had been named for various children of Tantalus, it was more or less luck that left Niobe as the namesake for this metal.

Some still call it Columbium, after Hatchett's name for it back in 1801.

Niobium is named after the daughter of Tantalus whose fourteen children were all murdered by Artemis and Apollo to punish her for boasting about her fecundity. A sad name for a not-sad metal.

90% of the niobium produced each year goes to steel alloys. The rest has a myriad of uses, such as MRI scanners, jet engine parts, superconductors and gas turbines. The glass used to make thinner lenses for the glasses of people with severe myopia is made with niobium. It is physiologically inert, so it makes good implant devices and hypoallergenic piercings; and it can be anodized into a beautiful rainbow of metallic colors, which makes it even more appealing as body jewelry.

Mystique

Rock on ancient queen
Follow those who pale
in your shadow

-Fleetwood Mac, *Gold Dust Woman*

When niobium is heated to various temperatures it anodizes into a rainbow of gorgeous colors. It is also very strong. It is thus prized by both body piercing enthusiasts and rocket scientists. It is easy to see why niobium has such an allure, or dare I say…mystique.

And what would such a multifaceted element do without an alias? Niobium has a second, less commonly used name: Columbium. Niobium is certainly a pretty appellation, but with a tragic backstory. In Greek mythology, Niobe is the archetypal grieving mother. All of her children were murdered by the gods. Why is this lovely metal named for this poor woman? Because she was a daughter of Tantalus, whose elemental namesake is in this group. In and of and as itself, niobium has nothing to do with grief. Unless one can grieve from being just too fascinating...

Mind you, this glamourous element does not hold itself apart from the herd. Most of its uses are in alloys. Its ability to keep its cool makes niobium a crucial component in jet and rocket nozzles. Its strength renders it invaluable in gas pipelines, and its superconducting skills are absolutely essential for MRI machines. It is not dependent on exclusion tactics to maintain its allure. It is poised in any company, as well as stunning when alone. Its charm is beyond question.

Niobium is here to give you permission to bring out your inner Emma Peel. You are tough, but beautiful. Mysterious, yet engaging. An excellent partner, and even better flying solo. Whatever you do today, whether you're getting groceries, fighting fires, or battling supervillains, do it with style. Pierce the heart of any problem, and trust yourself to stand the heat. You are gorgeous, rainbow and metal. Such is your mystique.

Molybdenum

Element Number:	42
Weight:	95.95u
Normal State in Nature:	solid
Melting Point:	2896° Kelvin
Boiling Point:	4912° Kelvin

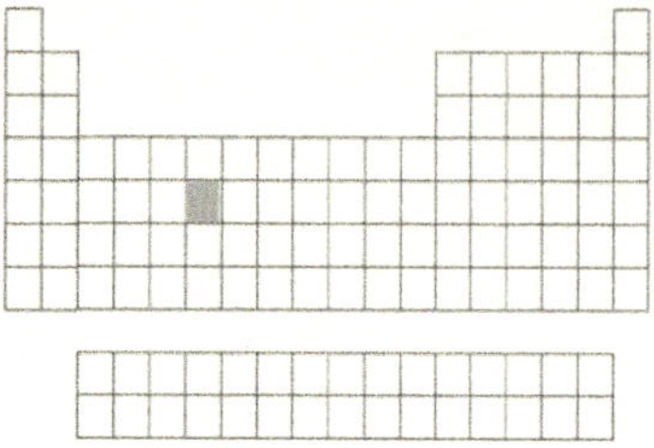

The ore called molybdena was historically mistaken for graphite or, sometimes, galena. It can be used to make marks on paper and lubricate things, and it was easy to assume its traits came from carbon or lead. In fact, the word *molybdena* comes from an ancient Greek word meaning "lead."

But in 1778 Scheele insisted that molybdena's features were caused by a new element, and Hjelm followed this up by isolating it in 1781. They named it after its ore.

It had very few uses for the next hundred years, although a Japanese sword dating to the 1300s has reportedly been found with a molybdenum steel blade. The technology to make such a thing is long lost. Its demand spiked during WWI, when it was used in armor plating.

Two mines in Colorado produce large amounts of the ore. Most of the resulting molybdenum is used in high-strength and high-speed metal alloys, as it is very heat resistant and tough. This is even true of its sulfur compound, used as a high temperature lubricant which does not seize under pressure.

Beyond that, it is a great fertilizer for cauliflower and some other crucifers. A very tiny amount in your body is essential. More than that can be toxic.

Moly is not a recreational drug, though. That's molly.

42

Wherewithal

"There's not a lot of meat on your claw, Crow."
"There's NO meat, Mike. I'm made of molybdenum."

-*Mystery Science Theater 3000*

Molybdenum hits its stride when it's under stress. As a metal, its strength is legendary at high pressure, and it can bear considerable heat. Industry is its milieu, where its alloys show up in high-speed high-stress tools, in pressure vessels, in gigantic factory motors; and its compounds show up as catalysts and even pigments, all of which have the wherewithal to handle immense force and intense temperatures.

Wherewithal is an innate ability to grit one's teeth and bear up under terrible pressure. We all have it, but we may not remember that we do. We may feel like we are ordinary transition metals, indistinguishable from all the others, until suddenly our strength is needed. We cannot test our hardihood during peaceful times; we must trust that it will be there when we need it. Like molybdenum.

Its most famous compound, molybdenum disulfide, is an industrial lubricant renowned for not seizing up under—this will sound familiar—intense pressure and heat. It seems that hardship is this element's calling. Humans are more complex and squishy, but we can surprise ourselves when a crisis happens, when we are suddenly given the job of dealing with unimaginable pressure. So often, against our own expectations, we can. Maybe not forever, and maybe we have to heal afterwards…But when the going gets tough, we find our wherewithal.

Whoa nelly, my friend. You appear to be dealing with some heavy, heavy stuff right now. If it feels like circumstances are pushing you from every direction, like you are being asked to withstand more heat and pressure than a person should reasonably be expected to bear, molybdenum has your back. Channel that horsepower, brace your feet, stiffen your spine, gird your loins. You can do this.

Technetium

Element Number:	43
Weight:	98u
Normal State in Nature:	synthetic solid
Melting Point:	2477° Kelvin
Boiling Point:	4538° Kelvin

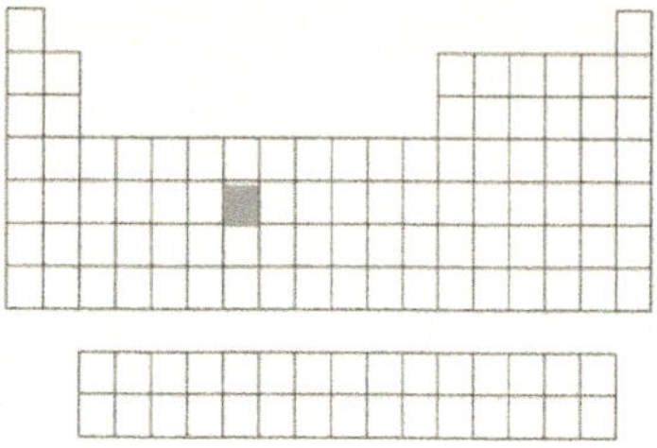

Element 43 was supposed to be low-hanging fruit for the scientists in the first half of the 20th century who were racing to discover and name every atom. It was first claimed to have been observed in 1925, but that was an error. There were lots of errors.

In 1936 Segrè visited Berkeley and convinced Lawrence, the inventor of the cyclotron, to let him take some old, discarded, radioactive parts home with him. Segrè and his colleague then proved that some of the radiation was actually coming from a new element, and it was the elusive 43.

In 1947, having established that there were no stable isotopes of this element and that it did not exist in nature, it was given the name technetium.

Technetium can be produced in red giants, but down here it's made from the waste material of fission rods in nuclear reactors, or *very* rarely as a spontaneous product of uranium decay.

The 99 isotope has a half-life of 211,000 years, but its metastable isomer half-life is only six hours. It has a ton of uses, almost all of them medical. At least 50 radiopharmaceuticals are based on it, as it works well with biochemical compounds and can therefore be used in functional imaging diagnostics. It is crucial in the early detection of bone cancer, among other things. It also has very little chemical toxicity, so the impact on the body is low.

43

Foresight

Then the cold and the storms and the darkness can do their worst. They can grope their way up the walls looking for a way in, but they won't find one, everything is shut, and you sit inside, laughing in your warmth and your solitude, for you have had foresight.

-Tove Jansson, *Moominvalley in November*

Mendeleev *knew* there was an element between molybdenum and ruthenium. He called it eka-manganese, but the space stayed empty for years. People thought they had found it several times, but their results were always impossible to replicate. Only a strong sense of foresight kept the space from being justified away.

When the element was finally found, it wasn't found at all; it was made. Against all odds, in the midst of stable, natural elements there sits an atom both radioactive and synthetic. It seems to be looking towards the higher numbers; a prediction of things to come.

Since its inception, technetium has been observed in tiny quantities in red giants, so it *does* have a natural place after all. But as far as humans are concerned, technetium is incidental, not integral. We make it. But why do we make a dangerously radioactive substance? Partly because it's one of the waste products left in a spent nuclear fuel rod, which is sad. But it is also produced on purpose, because it can be used in medical imaging. A six-hour half-life lets it leave the body in a hurry when a scan is finished, and thus do doctors catch bone cancer early.

Foresight predicted its existence, foresight is its role in the table itself, foresight is its purpose in medicine and doubtless foresight will help us find a use for radioactive waste. We must keep looking ahead.

Be prepared! Should be your motto, according to technetium. Give yourself plenty of supplies to do whatever might need doing. Buy toilet paper, apply sunscreen, pack extra sweaters, throw band-aids in your bag. Do NOT put off dealing with your emotional radioactive waste, and DO trust that what you are making space for today will arrive tomorrow.

Ruthenium

Element Number:	44
Weight:	101.07u
Normal State in Nature:	solid
Melting Point:	2610° Kelvin
Boiling Point:	4425° Kelvin

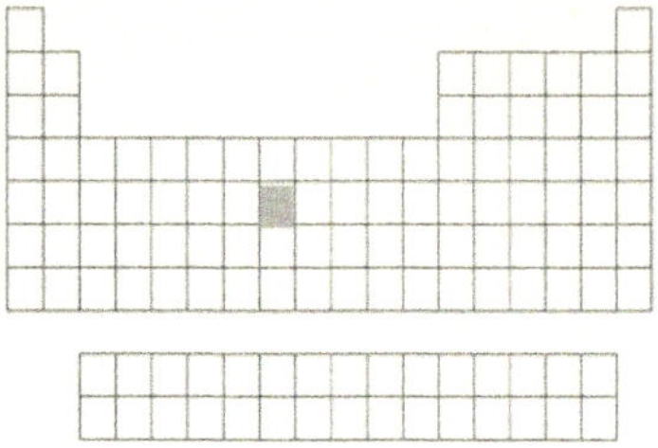

Based on atomic number, ruthenium is the first member of the platinum group, a set of six chemically inert metals that nestle together in the periodic table as well as being found in the same ores. Those ores were used by American first nations, and by Europeans in the 16th century, but it was not until the mid-1700s that platinum was recognized as an element, leaving the other components available to become identified.

The Russian Empire was rich in platinum ores. In purifying the platinum for coins, one of the residues was a different material, which was tricky to isolate and so was "discovered" more than once, by people who could not repeat their findings. Osann named it after the Latin word for Russia, but relinquished his claim on the finding. Claus rediscovered it and kept the name; the first element named for a country.

Most ruthenium still comes from the Ural Mountains of Russia. It is still, strictly speaking, a byproduct of platinum mining. It is used in alloys with other members of the platinum group, as a very small amount is all that is needed to make metal stronger. It is included in a superalloy to harden the turbines in jet engines, so, if you fly commercial airlines, it's your very good friend. The radioactive isotope is used to destroy melanoma in your eyes. It has a myriad of uses in electronics, always in conjunction with some other metal, usually platinum group members.

The catalytic converter in your car, which reduces pollutants in exhaust to a less toxic form, depends on the platinum group, ruthenium included.

44

Tribe

Belonging is the opposite of fitting in.

-Brene Brown

Ruthenium is the first metal of the platinum group, a small family of metals found together in natural ores as well as clumped together in the middle of the periodic table. They are all used in catalytic converters, but in addition to that each has its own unique properties. As with any tribe, the whole is more than the sum of its parts, but the parts are also more than a division of the whole.

Each of us has a tribe, whether of blood or of mutual esteem. We may share traits with them, but they do not define us. Rather, they support us, and we support them, because the purpose of a tribe is to protect and inspire; not to blend in with a crowd, but to be a member of a community. When we are separated from our tribe, or—worse—have yet to find our tribe, something vital in us starves. Connection is food for the soul.

Ruthenium is rarely separated from its platinum group family members. When it is isolated, its uses include electrical contact plating, because a thin film of it is sufficient to add a great deal of durability; and in thick-film chip resistors. As with people, the ability to facilitate and regulate communication is a skill to be prized, especially when flying solo. And we must also recall the entire group's use in catalytic converters. When a tribe works together, we can literally make the world a better place to breathe.

Ruthenium is here to tell you to unite with your people. Sisters, cousins, grandparents or grandchildren; or your writing group, your gaming pod, your high school buddies or your Rocky Horror castmates. Whichever of these lets you be authentic, inspired, and your most fundamental You...gather them. Be with them. Clear the air together, plug into reality together. Grow stronger together. Connect to your tribe.

Rhodium

Element Number:	45
Weight:	102.9055u
Normal State in Nature:	solid
Melting Point:	2236° Kelvin
Boiling Point:	3970° Kelvin

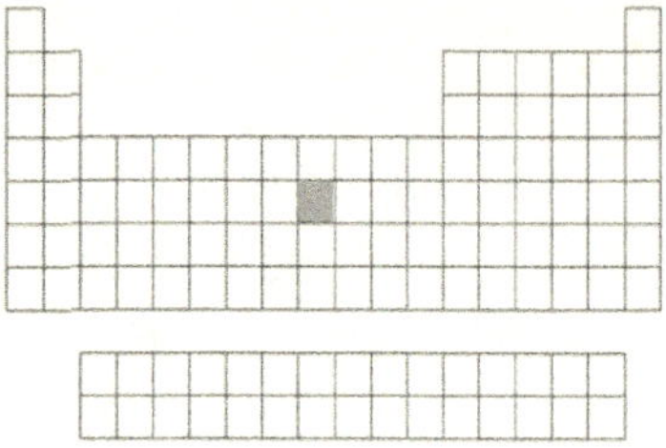

Rhodium was identified by Wollaston from the same South American ore in which he discovered palladium. He named it for the Greek word for "rose," because its precipitate was a lovely shade of rosey red.

It is exceptionally rare and expensive, so for a long time its uses were few. It was added to thermocouples because it is unusually stable at high temperatures. Eventually it was used in electroplating, because it is so shiny and reflective. Finally it was included with the other platinum group metals in catalytic converters, and this is its biggest market to this day.

It is still one of the rarest metals, and shiniest. Aside from its use as a catalyst, rhodium can be found in jewelry plating, thermocouple elements and electrical contacts.

Paul McCartney has a rhodium-plated decorative record on his wall, because he has surpassed mere silver, gold and platinum.

Considered a noble metal, pure rhodium is inert in the body (not that you will ever have the opportunity to consume enough to poison yourself, regardless) but some of its compounds are toxic. Also, you don't want to breathe it. That goes for most metals.

45

Playfulness

Every man should dance for life. Not being a dancer, but dancing.

- Rudolf Nureyev

Rhodium is ever so shiny, ever so durable and ever so precious. Named for the rose-colored chlorine compound used in its discovery, it is incredibly rare and, because of its chemical inertness, even considered noble. So what massively important task do we ask of this strange and elusive member of the platinum group? Why, plating costume jewelry, of course!

Okay, there are other and more serious uses for rhodium, including coating electrical contacts and of course catalytic converters. It can thrive under high temperatures and in high traffic. But, oh, it is so shiny! Even shinier than chrome! It can scrub the air with its family, yes. But it can also dance in its own reflection and be silly for a while. It knows the importance of playfulness. To have the ability to do big, important things, and to instead focus on fun and frivolity, is an underappreciated skill. It freshens our perspective. The daily grind of being an adult in this world will wear us all down, but when we play, we prevent that. To be playful is to heal.

Playfulness, like rhodium, can be a rare commodity. We fall into the trap of taking the world so seriously, we forget how to smile. It is crucial to remember that our inner child still needs care and attention…and shiny shiny dress up time.

Rhodium acknowledges the importance of your work, the seriousness of your calling and the weight of the world on your shoulders. Now, put all of that aside for a while and have some fun! Put on a costume, go romp in a garden, blow bubbles in the sunshine or have a five-year-old paint your toenails while you watch Looney Tunes. When you go back to work, bring this sense of playfulness with you, and you may find that your tasks are lighter, shinier, and easier. Keep playfulness in your toolbox.

Palladium

Element Number:	46
Weight:	106.42u
Normal State in Nature:	solid
Melting Point:	1825° Kelvin
Boiling Point:	3240° Kelvin

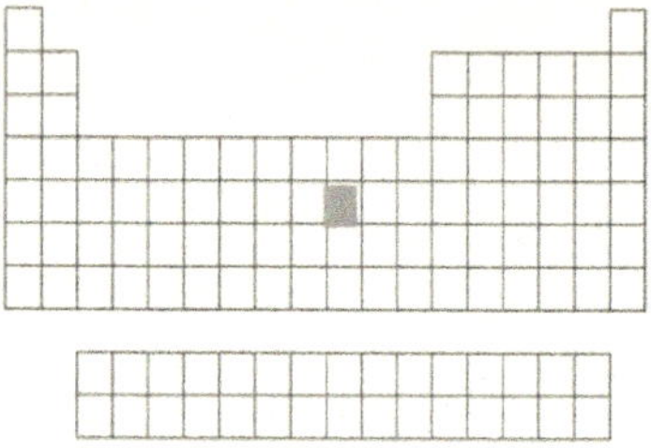

Wollaston discovered, purified and named palladium in 1802, three years before he went public with it. It is hard to say why he waited. Perhaps he was just being whimsical. The reason he chose that name, at least, is clear. According to his notebooks, he discovered the new noble metal in July; The asteroid Pallas had just been discovered the previous spring.

Pallas, by the way, was the giantess in Greek mythology who was raised alongside Athena as her foster-sister and sparring partner, who was killed by accident during one of their mock fights. Athena, grieved, added her friend's name to her own, and had a statue built in her honor. Since then, Pallas Athena (or even just Pallas) is a reference to the goddess; whether in corvid poetry or in asteroid naming.

Palladium, like the other platinum group metals, is invaluable in catalytic converters. In fact, it facilitates all kinds of chemical reactions, and is the go-to catalyst for a long list of important processes.

Beyond that, it is used in jewelry plating. "White gold" contains a palladium alloy, as it is a pale-colored metal. Sometimes it is used in dental fillings, although it has been known to cause allergic reactions in a small number of people.

One interesting trait is its ability to absorb, purify, and store hydrogen at normal pressure and temperature. Several transition metals can do this, but only palladium can take on 900 times its own volume in hydrogen. It is impressive.

Acceptance

Do not cast doubt upon that of what I am nor whom I shall serve.

-Kuiil, *The Mandalorian*

Meet the element named for an asteroid that was named for a goddess who was nicknamed for a giant she'd accidentally killed. Would we accept such a convoluted reason for our own appellations? Palladium does. Palladium is the member of the platinum group with the lowest melting point and the lowest density. Although it is certainly a precious metal, and costs an arm and a leg, palladium is a very accommodating character. Its uses are all over the map, from jewelry plating (it is even shinier than rhodium!) to dentistry, groundwater treatment, and the catalytic converters which keep it, and the rest of the platinum group, in high demand. Our ability to accept a wide range of tasks and carry them out is one of the most precious things about us, too.

But palladium has a trick that its siblings do not share. It can absorb hydrogen at an astonishing rate, at normal pressure. Because of a quirk in its crystalline structure, up to 900 times its volume in hydrogen it can quietly soak up, and even purify. To be that accepting of source energy is perhaps the most incredible feat of a very impressive metal. And you may be surprised to learn that we have it, too. We can accept strength as well as hardship; gifts as well as duties. We can accept a lot more than we think we can.

Say Yes today. Say yes to the things you wished for but thought you could not have. Say yes to challenges and adventures and your own emotions and especially to gifts. If you have been accepting neglect or abuse, stop that. You are too precious a metal for that. No, accept things like...power over your time and your boundaries. The process of healing or the knowledge that gives closure. Yes to all of it. Alloy with those who value you, and tap into the source whenever you need to. You are splendid. Accept it.

Silver

Element Number:	47
Weight:	107.8682u
Normal State in Nature:	solid
Melting Point:	1235.08° Kelvin
Boiling Point:	2436° Kelvin

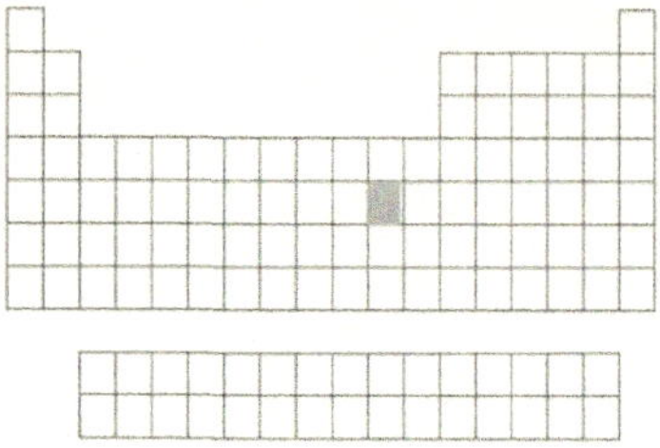

Silver is one of the oldest metals known to be used by humans. Artifacts more than six thousand years old have been found in tombs, especially jewelry and religious articles. It's been used as a currency for most of that time. The rise of Athens can be attributed to a nearby silver mine; the stability of Roman Empire currency was reliant on this metal.

The word for silver is surprisingly similar between Old English, Norse, German and even Russian languages: Seolfor, sølv, silber, serebro, you get the idea.

In Latin, though, it was called *argentum*, meaning white or shiny, which is where the elemental symbol comes from. New World silver mines became so important, they are where the name of Argentina came from.

Silver is extremely ductile; it can be drawn out into a wire one atom wide. It is also malleable, thermally conductive, highly reflective and electrically conductive.

Silver is often an incidental finding in lead mining, which is the major source today. It has been widely used in medicine and health care, from infant eye drops to burn care, and antibacterial coatings for bathroom faucets. It has more uses than can comfortably be discussed here, from catalysis to oldschool photography to forks. Nearly every religion has silver holy objects. It is widely used in electronics, werewolf and vampire defense, and even in baking.

Eat enough, and you'll turn blue-gray.

47

Clarity

"The rose," he announced. "Loveliest and most formidable of flowers. Arms of York and Lancaster. In medieval times, symbol of Jesus. Always, it has meant, beauty, love, peace…" He presented the bud to Martha Macnamara. It lay resting on his long fingers until she scooped it up. She sniffed it and held it up to the light. "Symbol? What's a symbol? This is a rose." She smiled and walked on.

-R.A. MacAvoy, *Tea With the Black Dragon*

Science uses silver mirrors when it needs an absolutely flawless reflection. Engineers use silver in circuit boards where a totally reliable connection is required. Silver has the distinction of being the most reflective and the most electrically conductive element in the universe. Reflection and communication—these things are at their best when clarity is prioritized.

Silver is also the most thermally conductive. The heat of human emotion becomes a grace rather than a burden when it is given space in which to *be*. That is one role of clarity: To allow. It does not require explanations or excuses. Clarity, like a swiftly flowing stream, exists beyond all words, in the place where the mind can embrace experience without needing to justify it.

Silver does not remain untouched by the world it reflects and channels. It can oxidize over time. It needs to be maintained. Even when black with tarnish, its conductivity is preserved; but that is not an excuse to leave our sense of clarity to look after itself. To know clarity is an act of courage and surrender both; to choose it every day is to strengthen it; to strengthen it is to strengthen our souls.

A reminder you may sometimes need: You are, at heart, a being of absolute clarity. Look around and look within; see what is there without filters or judgment. There is no action required here, no map, no instructions beyond this: Allow yourself to be what you are. Allow the universe to be what it is. See it with fresh eyes every moment, for it is fresh every moment. Reflect with purity, communicate with transparency, experience feelings with lucidity. You are silver.

Cadmium

Element Number:	48
Weight:	112.411u
Normal State in Nature:	solid
Melting Point:	594.26° Kelvin
Boiling Point:	1040° Kelvin

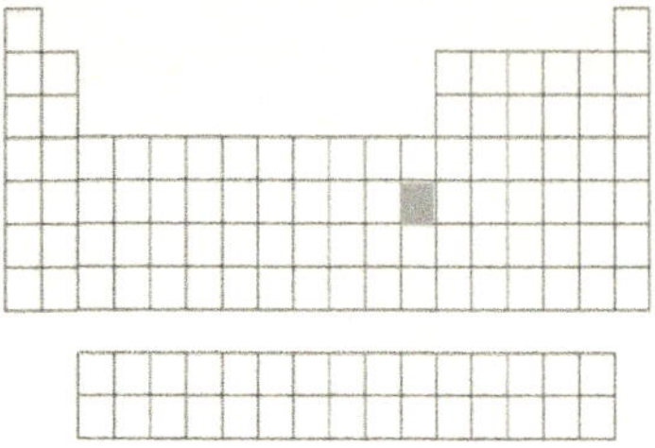

Calamine is a mixture of minerals (primarily containing zinc), named after Cadmus, who founded Thebes. The Latin name for this mineral is where cadmium gets its name, because cadmium was discovered as an impurity in this ore.

Two German chemists, Stromeyer and Hermann, simultaneously discovered cadmium in 1817. For a century afterwards, Germany was the primary producer of the metal.

Cadmium's red spectral line was used to define the angstrom from 1907 to 1960.

Cadmium's potential as a pigment was seen as early as the 1840s, but it wasn't until the 1950s that production of cadmium red, orange and yellow really came into play. The other major use for it in the first half of the 20th century was in coating steel, to make it resistant to corrosion.

The pigments are still available, and are incidentally so bright they have to be muted before use in painting. Unfortunately, cadmium is a heavy metal, which builds up in environments and in bodies and causes a lot of damage. The symptoms are most often seen in post-menopausal women, causing kidney failure and heart disease. Tobacco smokers inhale cadmium molecules in the smoke.

The 1980s and '90s slapped a ton of regulations on cadmium, and these days most of it is tied up in batteries, and in nuclear reactor control rods as a neutron poison.

Admire it from a distance.

Challenge

"I figured out a process! It's just violence."

-Simone Giertz

Cadmium is a heavy metal with an aggressive personality. For a long time, it has been used as a key component in batteries, but its toxicity makes that a short-sighted idea. It is not the harm it does to ecosystems that makes it synonymous with Challenge, though; plenty of elements are poisonous. But few of them are so very aggressive, bold and uncompromising in all endeavors.

The brightest red and yellow pigments are cadmium compounds. Steel airplane parts are plated with cadmium to halt corrosion. Control rods in nuclear reactors use cadmium, because it will smack down a cascading nuclear reaction and stop it in its tracks. Science is even developing medical uses for it, hoping it will have a similar effect on cancer cells. Wherever it goes, cadmium is kicking ass, challenging the universe to Bring It On, insisting on work, calling attention, commanding respect…demanding a challenge. There are worse ways to exist in this world.

But one must never forget the batteries. Cadmium's weakness is a willingness to pump out an enormous amount of energy today, only to leach poison into the water table tomorrow. We are responsible for our consequences; we must rise to that challenge.

Take on all comers, my friend. Taunt the universe from a rooftop. Strap on your boxing gloves, derby skates, F1 helmet, or power tie, and do the thing that you thought was beyond your ken. In fact, don't just DO it; punch it. Attack the task until sweat is flying off you and people are backing away in awe. You got this. You OWN this. The world has never seen what you're bringing today. Your colors are hot and your energy is off the charts.

Then recharge those batteries. Your inner power will turn toxic if you neglect it.

Indium

Element Number:	49
Weight:	114.818u
Normal State in Nature:	solid
Melting Point:	429.78° Kelvin
Boiling Point:	2350° Kelvin

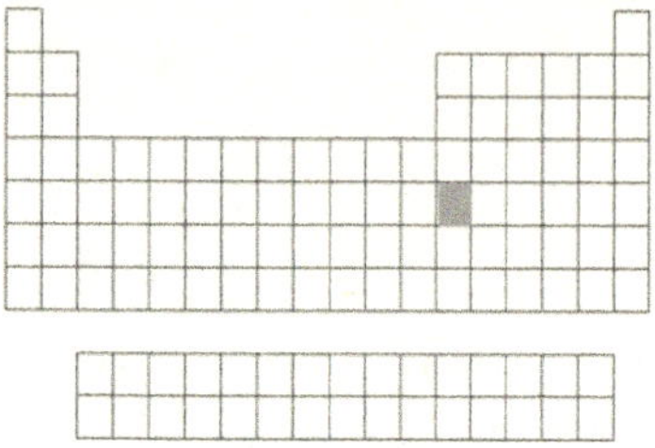

Reich and Richter were an excellent chemistry team, not least because Reich was colorblind and needed Richter to describe to him the colors of the spectral lines in their experiments. While hunting for what they thought was arsenic in some ores from Saxony in 1863, they spotted a strange blue emission line that was not associated with any known element, and decided to investigate. They named the hypothetical new element indium, after that telltale line of deep blue.

A year later, Richter isolated the metal, claimed the discovery as his alone, and alienated his old lab partner forever. Hubris is a terrible thing.

Indium is a metal, but so soft you can cut it with a knife, and draw with it on paper. Its uses are mainly in electronics and nuclear reactors. It makes alloys soft; it is used in solder, semiconductors and thin film LCDs. Indium wire is used as a vacuum seal in ultra-high-vacuum applications, and its high cross-section for thermal neutrons gives it a job in nuclear control rods. It also makes glass wet.

Mixed with yttrium and manganese, it makes a vibrant blue pigment.

When you bend it, it cries.

Don't breathe the dust; it'll give you indium lung. Which only about ten people have ever officially had, so you're probably safe.

Lassitude

"You are what you eat."
"I am a Fritos corn chip."

-Nicole Hollander, *Sylvia*

What sets indium apart is how soft it is. It is a metal, yes; but where the other metals celebrate being hard, tough and rugged (or in the case of alkali metals, at least explosive), indium is just bendy. It gives no shits. It puts forth no effort. Even the indigo spectral emission line it was named for, is its *only* spectral emission line. If the metals were humans, the whole D block would be the cast of Top Gun, and indium would be Dorothy Parker.

Still, even the element of apathy has plenty of jobs to do. It makes a very effective seal on high vacuum chambers; it lends its languor to solder and semiconductors. It is even a critical component of LCD TV screens, as one of its compounds (ITO) is uniquely both transparent and conductive.

This is important. Even at our most ennui-ridden, we can contribute to our world. Even if all we're doing is being lucid, or holding a space, or just plain melting…sometimes that is what is called for. We can whine when the universe needs us—and indium, like tin, does in fact make a crying sound when bent—but it's not as if we can't accomplish plenty just by lying here for a while. No sweat.

Good news if you're feeling lazy! Indium wants you to chill. Lie down in a comfy place, staring at nothing, and exist for a while. If you get bored and want something to do, go find an energy vacuum and seal it. Or a metaphorical leaky pipe that needs your spiritual solder. Or try communicating, as you are particularly limpid right now. But if all you want is to just take a stinkin' break from it all and check out for a while, do it. This is your lassitude card.

Tin

Element Number:	50
Weight:	118.71u
Normal State in Nature:	solid
Melting Point:	505.12° Kelvin
Boiling Point:	2876° Kelvin

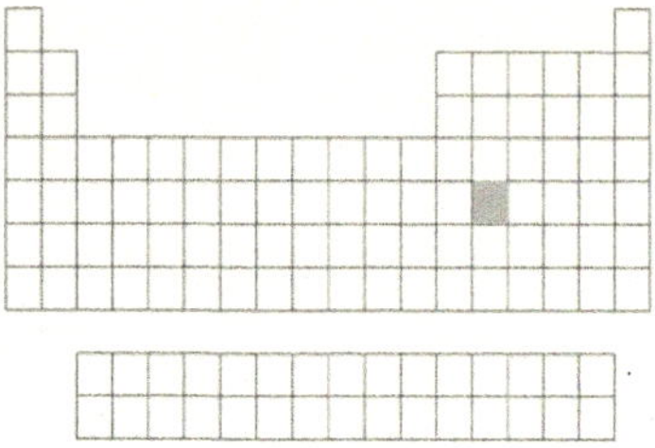

When tin was first alloyed with copper, the Bronze Age began. Tin makes copper melt at a lower temperature, flow better when molten, and harden to a stronger metal. Bronze artifacts can be beautifully detailed because if these qualities, and even now cast bronze sculptures are made using processes that have changed very little over the centuries.

Tin-containing ore is not available in very many places, so, when alloying technology became widespread, trade routes were established. Smaller, more isolated cultures started to intermingle as a result of the advent of the Bronze Age. In a way, tin brought the ancient world together.

Tin cries when you bend it. Much of the tin produced is used in lead-free solder, which has applications in plumbing and in circuitry. It is still used in bronze; its alloy with antimony, bismuth and silver gives us pewter; and it even pairs with niobium (superconducting magnets are so cool). Tin does not corrode readily, so it is frequently used to plate steel food containers and copper cookware. Both steel and copper are vulnerable to the acids present in, for example, tomatoes; tin is not.

The Earth is estimated to run out of mineable tin sometime around 2060. However, it is recyclable, and much of its current production is already through recycling.

The sound it makes when you bend it is described as "tinny." Aha!

50

Flexibility

"I can believe things that are true and I can believe things that aren't true and I can believe things where nobody knows if they're true or not…I believe that the future sucks and I believe that the future rocks and I believe that one day White Buffalo Woman is going to come back and kick everyone's ass…"

-Samantha Black Crow, *American Gods*

Easy to melt, easy to cast and nontoxic, tin has always been a fan favorite. Paired with copper, it gave us the Bronze Age. The Roman Empire used it to make flatwear, and to this day it lends its skill to things ranging from food packaging to optoelectronics. Flexibility is its prime trait. One can bend a bar of it without much effort, though it will cry aloud as the crystalline structure reforms itself. We, too, may complain when we're forced into new positions, but that's fine. Flexibility does not imply, or demand, silence.

Tin has two allotropic states. Alpha tin is a dark grey powder, basically useless. Heat that, and it melts and coalesces into beta-tin, which we described above…but if left in the cold, "tin pest" will turn it back into dust. Ashes to ashes, tin to tin…what has changed will change back again. One must be flexible.

We, too, can harness flexibility. Because whatever happens, whatever we are asked to do, and especially when we have been cold so long we turn into dust…there will always be another opportunity to melt, to reform, to change and to grow. It may make us cry out, but it will never break us.

Time to bend, my dear. Perhaps someone else's needs are a thing you are going to have to incorporate into your life; perhaps a job loss or medical problem or promotion or recovery has got you feeling like you're the wrong shape for your space right now. You have it in you to modify exactly as much as you need to, to make yourself comfortable again. Take a stab at it. Make the necessary changes. Bitch as much as you need to. And stay warm.

Antimony

Element Number:	51
Weight:	121.76u
Normal State in Nature:	solid
Melting Point:	903.91° Kelvin
Boiling Point:	1860° Kelvin

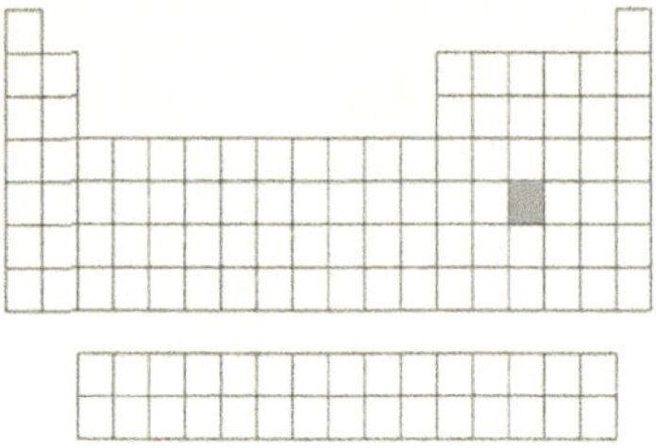

Antimony has been around for a long time. Even disregarding the fact that antimony sulfide was used in ancient Egypt as a cosmetic (kohl), and used in ancient Rome as a treatment for various ailments, humanity has also had the ability to smelt it into its metallic form for time out of mind.

Alchemists through the ages have been familiar with antimony, but it's unlikely any of them could tell us where its name comes from. Some believe it comes from *anti-monachos*, meaning "monk-killer," because alchemists were frequently monks and antimony is toxic. But it's not all *that* toxic, so the origin is probably deeper and stranger than that. At least we know it was Berzelius who gave it its abbreviation, Sb, stemming from *stibium*, which comes from an old, old name for the eyeliner.

Uses for antimony include lead alloys for bullets, ships' keels, and old-fashioned movable type. In fact, early printing presses would have been much less feasable without antimony. It expands slightly as it cools, so type blocks cast with it were very sharp and durable. And, speaking of archaic applications, wine used to be served in antimony goblets by doctors who, for whatever reason, wanted their patients to throw up.

More than half of antimony today is used as a flame retardant on children's clothing and toys, and in car and airplane seats. Gone are the days of goblets, and only a few places still use kohl.

Bullets, however, aren't going anywhere.

Resonance

There is this thing that's like talking except you don't talk
You sing
You sing…

-Dresden Dolls, *Sing*

Antimony is a hard and brittle metalloid, poised between the definitely metals and the definitely not. Long ago, Egyptians used it in cosmetics. Today, it has fireproofing applications. In between, it has had a wide-ranging career. A chord is more than one note, and so is antimony. Its brittleness makes it difficult to work with, but its hardness has found many uses; mainly alloyed with lead, which is too soft for several of its tasks. Bullets require a smidgen of antimony; and Gutenberg's printing press would have been impossible without it. The moveable type was made with an antimony/lead/tin alloy that turned out sharp, reliable and long-lasting letters. What a resounding legacy! Antimony is also toxic, and was for a long time used in medicine to induce vomiting. This is gross, sure; but music can also be jarring. Different jobs in different eras.

Most metals make pinging or crackling noises as they cool from molten to solid, but antimony makes a sound like chimes, haunting and melodic. Resonance, more than any other one thing, is the voice of antimony. It vibrates to whatever tune is playing, and whether that means being smeared on eyelids, or hardening bullets, or bringing literacy to medieval Europe, antimony is here for it. Singing.

It is time to sing. Sing the songs of your childhood, or a song you wrote yourself. Sing your friends' favorite songs and the songs from your most cherished Broadway musical. Sing along with the radio in the car and in the kitchen. You don't have to be good at it. Your voice, once raised in song, is going to grow stronger, and you will be heard more profoundly in all else. Your eyes will be bolder, your typeface will be clearer. You will bring strength where strength is needed. But first, you must sing.

Tellurium

Element Number:	52
Weight:	127.6u
Normal State in Nature:	solid
Melting Point:	722.72° Kelvin
Boiling Point:	1261° Kelvin

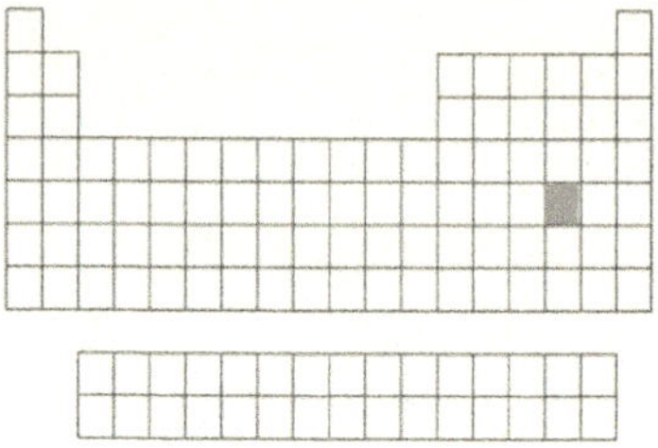

Von Reichenstein discovered tellurium-bearing ores while visiting a Transylvanian gold mine in 1782. He spent an exhaustive three years doing more than fifty tests to discover whether he had found bismuth or antimony or whatever; and finally concluded that he now knew everything about this metal except for what it actually *was*. He named it *metallum problematicum* and gave up.

It was then left to Klaproth in 1798 to find it again, declare it a new element and name it tellurium, meaning Earth.

As an additive to gasoline, it prevents engine knocking. But since it smells like rotting garlic, that application was never pursued.

Tellurium is used in metallurgy, as it improves the machinability and strength of a number of different alloys. As a chalcogen, it is closely related to selenium, as well as (to a lesser degree) sulfur, polonium and oxygen.

It shows promise as a material in solar panels because of its electronegativity. However, it may be too rare, and its mining and refinement too ecologically ruinous. We shall see. Another use that has fallen by the wayside is rewritable CDs, which use tellurium suboxide in their rewritable layers.

Rapidly changing technology may or may not result in higher demand for tellurium, for applications as yet uninvented.

Insouciance

I won't worry and I won't fret
Ain't no law against it yet

-Cyndi Lauper, *She Bop*

What can we learn from Tellurium? What are its lessons in this wonderful world? It is one of the rarest elements on Earth, yet its name means "Earth." How do we interpret that? Or should we just chuckle and move on?

Tellurium is considered technology critical, as it is used in DVDs and *some* solar cells; but those uses may soon be obsolete. Or maybe not. Who knows? Its use within these applications is based on the fact that its reflectivity changes when a laser is bounced off its surface, so it's amusingly meta that tellurium is either cutting-edge OR a relic, depending on how you look at it.

Insouciance is an interesting word, because its definitions range from apathy or disinterest to ease and cleverness. As a concept, it is dependent on context and application and probably time period. Which is why it's the perfect word to apply to tellurium, who is easygoing! Or possibly indifferent. Or stoic or phlegmatic. Depending on the lasers.

It seems like the sort of element that ought to take itself seriously—after all, it is a chalcogen, and those guys are major movers and shakers—but Tellurium doesn't do that. Instead, it is smelly! Rotten garlic smelly. As an isolated element it stinks, and tellurium toxicity will cause a person's breath to reek for days. Not fun, but a little bit funny. One more thing to chuckle about.

When you have drawn Tellurium, you should decide exactly how you are not going to give a crap. This world makes a lot of demands on you, and you have to just let some things go. Release them. Reflect them. Are you going to do this with sadness? Listlessness? Relief? Laughter? Torpor? Affability? All of these are available to you, and now is a good time to pick one. Or two. Change it up. Be shiny and *smelly. Be Tellurium.*

Iodine

Element Number:	53
Weight:	126.90447u
Normal State in Nature:	solid
Melting Point:	386.7° Kelvin
Boiling Point:	457.5° Kelvin

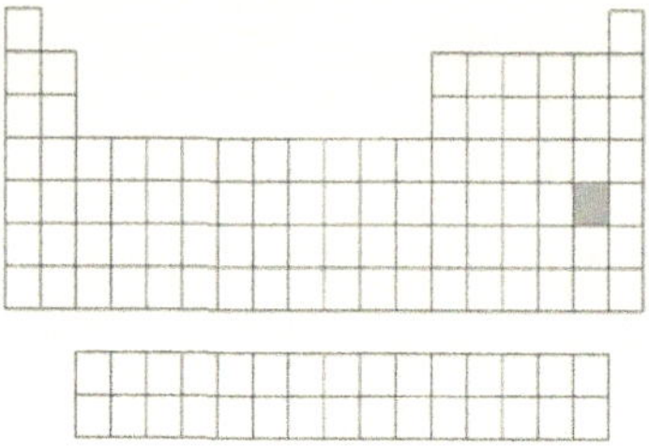

Courtois was the son of a saltpetre producer, trained by his father in the complex process, and he did it well...until one day in 1811 when he added too much sulfuric acid to the seaweed ashes and saw a cloud of purple vapor as a result. Lacking the resources to chase what he thought was probably a new element, he sent some of the crystallized vapor to several of his chemist friends. One of them, Gay-Lussac, suggested the name *iode* from the Greek for "violet." Another, Davy, confirmed that it was chemically similar to chlorine. Those two argued viciously over who had identified iodine first, but they did both agree that Courtois was the first to isolate it.

Iodine's powerful antiseptic properties were unlocked in 1873, and first used in a surgery in 1908.

It is still used to sterilize skin before surgery, and to block the thyroid gland from absorbing radioactive iodine produced by nuclear explosions. In fact, many thyroid problems are treatable with iodine.

This is because iodine is a vital nutrient, used by the body to synthesize two important growth hormones. It is especially critical for children to get enough in their diet. A deficiency can be dire. Table salt is iodized for this reason. It is also used in livestock feed and medicines, and in spacecraft propulsion.

Undiluted, however, it is toxic. It denatures proteins, including enzymes. This is not something you want happening inside your body.

53

Shield

And this is what cosplay is about. Agony! and pain! And it's so much fun!
-Svetlana Quindt, *Kamui Cosplay (The Hylian Shield video)*

For some of us, our direct experience of iodine involves waking up after a surgery with orange stuff smeared across the relevant body part. Iodine is a mineral, but once dissolved in a solution it can be used as a potent antiseptic. It wards off infections when the stakes are high. Those who have horses are also likely to recognize iodine as the substance that keeps hoof rot at bay. Again, high stakes, and again a defense, a deflection. A shield.

But a shield is not only useful in a melee; it is also effective against ranged weapons. And iodine isn't just for surgical prep and hoof health, but also for day-to-day wellbeing. It is the heaviest of the mineral micronutrients, and our bodies use it constantly. Without it, a person can succumb to hormone imbalances, loss of intellect and goiter. We don't see a lot of people with goiters in our culture, largely because of our iodized table salt. We are shielded.

Radioactive iodine is a major component of nuclear fallout, and people who have to work in such areas are at terrible risk of assimilating the bad iodine and getting sick. But there is a way to minimize danger: Take a really large dose of stable iodine before going in. The thyroid will not be able to absorb the nasty stuff when its iodine receptors are already full. So, hoofs and surgeries and goiters aside, iodine is a shield so strong it can even defend against…iodine. Well done.

Get those defenses up! Strengthen your wards and see to your hooves. Cleanse your space and cut off parasites. There may be something painful approaching, and you must make sure your personal boundaries are in good form before it does. Then again, there may be no storm coming, but iodine is here to remind you that your daily wellness must not be neglected. Support your own infrastructure. Pay heed to your hormones. Raise your shield.

Xenon

Element Number:	54
Weight:	131.293u
Normal State in Nature:	gas
Melting Point:	161.39° Kelvin
Boiling Point:	165.1° Kelvin

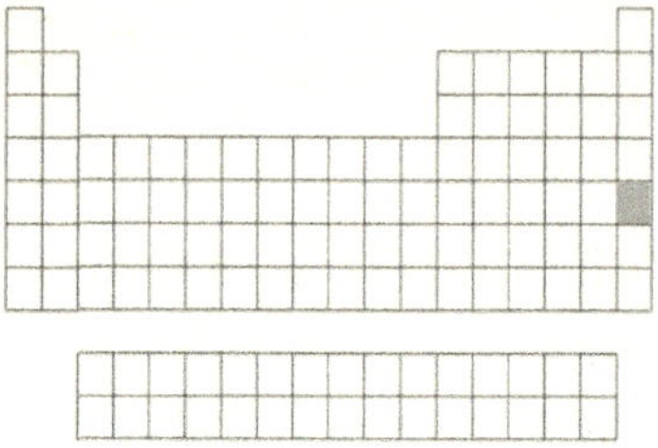

Xenon was discovered by Ramsay and Travers in 1898, the same team that had already found neon and krypton. They named this new noble gas "xenon" from the Greek for "stranger."

In the 1930s, Edgerton invented the xenon flash lamp that he used for high-speed photography. Meanwhile, Behnke was studying the unpleasant effects of deep-sea diving. In playing with the mixture of gases used by divers, he accidentally discovered xenon's properties as an anesthetic. After a careful decade of study, in 1951 it was first used for that purpose in a human surgery.

Though most noble gases do not interact with other elements, except under extreme duress, xenon was the first to be found to actually form a compound.

You can't see or smell xenon. It is still used in flash and strobe lighting for photography; a similar device makes xenon useful in lasers. It has very low thermal conductivity and low ionization potential, so it's the perfect starter gas in sodium streetlights.

Though expensive, xenon is as yet one of the most impressive anesthetics, with fewer side effects than the conventional kind. It keeps nerve cells from getting damaged when they can't get oxygen; it has neuroprotective properties and is even considered environmentally friendly, as (unlike nitrous oxide) it is not a greenhouse gas.

Its oxides make big explosions.

It has potential in ion propulsion for space travel.

54

Strangeness

You'll like the boy. He's as weird as snake's suspenders but sweet as a stolen kiss.

-Jubal Harshaw, *Stranger in a Strange Land*

When it comes to stolen kisses, xenon should come to mind. Alone in this family of completely inert noble gases, xenon does, on occasion, form bonds with other elements. Mostly fluorine, the crazy anarchist. There is a certain pleasing harmony to that. Our xenon is not afraid to let its freak flag fly.

In the main, though, xenon is still inert, and used for various things; as the high-pressure gas in movie theater arc lamp projectors, and even as a general anesthetic. Two ways to set a person dreaming, both dependent on this eccentric gas. But wait! There's more! Xenon has several radioactive isotopes, the cosmic presence of which has made it possible for scientists to determine the age of our solar system. Radioactive decay is a frightening and destructive phenomenon, but useful for knowledge-seekers as well. How very…strange.

All of this is to show that being a bit…bizarre, so to speak, is a blessing. It brings light to dark places and knowledge to those willing to learn. Strangeness is an asset the world could use more of.

Hooray! You have drawn xenon! It is time to get weird.

Tear off those shackles and free your inner oddball. Find someone fearless to connect with. If you have been still, get up and boogie; if you have been silent, start quacking. Do not allow embarrassment to slow you down; keep in mind that your weirdness shines like a beacon, giving relief and even joy to those who need it. Anyone who matters would rather join you than judge you. Be strange.

Cesium

Element Number:	55
Weight:	132.90545u
Normal State in Nature:	liquid
Melting Point:	301.54° Kelvin
Boiling Point:	944° Kelvin

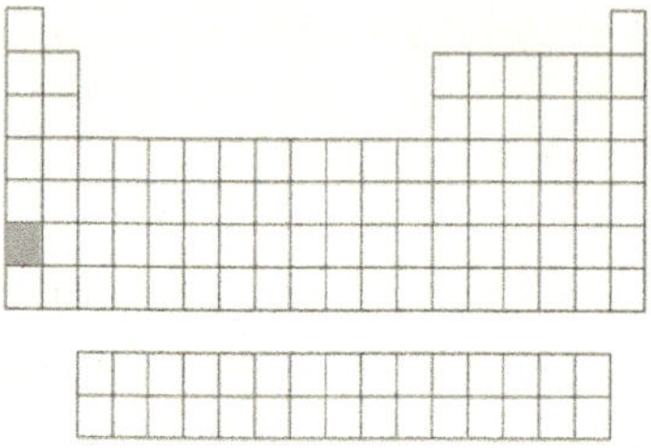

Bunsen and Kirchhoff, having invented the spectroscope, put it to good use by looking for the emission signatures of elements in a variety of matter. A light blue line found in mineral water led them to the discovery of cesium in 1860, and the name comes from that line (*caesius* is Latin for sky blue). It took a lot of work. Reportedly they used 44,000 liters of the water in question to yield them, among other elements, 7.3 grams of cesium chloride. They did not manage to isolate it; that was done in 1882 by Setterberg.

A "getter" is a material you put inside of a vacuum, to sequester any oxygen that may sneak in, and cesium has been used as a getter since radio vacuum tubes were first produced in the 1920s. Its functionality in a ton of other fields was explored in the 1950s. Some of these are still used today.

It was in 1955 that the first accurate cesium clock was built. The methods have been refined over the decades, and these days their rate of error is 1 part in 10^{15}. The Cesium Standard is what all our tech is linked to. The second is defined by cesium.

Aqueous solutions of cesium make an environmentally friendly lubricant for oil well drill bits. This is its largest current application. A similar solution is used as a centrifugation fluid in labs.

Being an alkali metal, it is extremely pyrophoric, explosively releasing lavender flames when exposed to even to slightest hint of moisture. Its melting point is a high room temperature (83°F/28°C) so is represented here as a liquid.

Precision

Today's forecast: cloudy with a 95% chance of immaculate dark academia vibes.
-Bernadette Banner

The second is defined as being equal to the time duration of 9,192,631,770 periods of the radiation corresponding to the transition between the two hyperfine levels of the fundamental unperturbed groundstate of the cesium-133 atom.

Forgive the jargon; all this is to say: Cesium is used to measure time. There are 300 cesium clocks around the world, kept synchronized. They are so accurate, they will lose less than a second in the next 20 million years. As if that wasn't enough, cesium is needed in particle physics research. When paired with a halogen, it can detect tiny amounts of gamma and x-ray radiation, or minute fluctuations in magnetic fields. If one is to be precise, one must start with being perceptive.

But there is more to cesium than measuring intangibles. When dissolved in a solution, it creates a fluid dense enough to float rocks. This can be useful in microbiology labs, as a centrifugation liquid that can isolate viral particles or nucleic acids. It is also used to literally float rocks out of mineshafts. And finally, of course, a warning: Cesium is the most reactive of the available alkali metals. It ignites spontaneously when exposed to air, and dropping it in water will cause an explosion notable for being instantaneous rather than impressive. When one is precise, one can also be very, very swift.

Cesium is telling you to practice precision. Pay attention to the details; tighten up slop and sharpen your focus. Make sure your numbers are exact, your lines are clean, your seams are straight. If your day also involves the equivalent of floating rocks, bring to that the same care and attention as if you're measuring the vibration of electrons. Keep you perception open, stay dry if you don't want to explode, and most importantly, remember: Time does not define you. Quite the opposite, in fact.

Barium

Element Number:	56
Weight:	137.327u
Normal State in Nature:	solid
Melting Point:	1002° Kelvin
Boiling Point:	2078° Kelvin

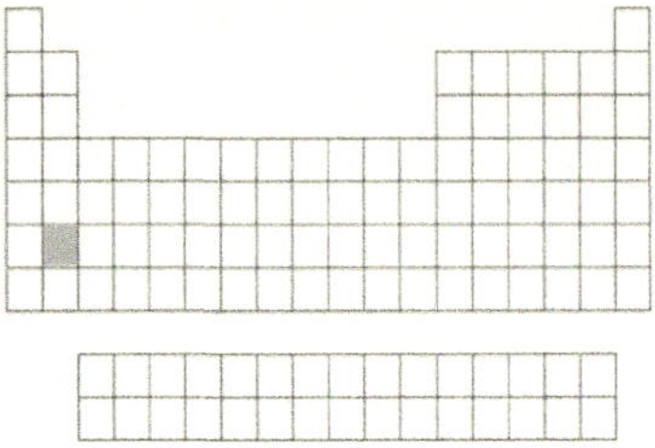

An Italian cobbler-alchemist seeking his fortune and the Philosopher's Stone near Bologna in 1602 found a pebble of impure baryte that, once heated in a brazier, became a glow-in-the-dark material. It transpired that many of the stones in that region had the same property, and "bologna stones" became a prized commodity among alchemists.

In the 1770s, both Scheele and Gahn hypothesized that there was a new element in baryte, but neither succeeded in isolating it any further than its oxide. Finally, in 1808, Davy cracked the code and isolated this alkali earth metal, naming it for the mineral.

The x-ray machine was invented in 1895; the first time barium was used as an x-ray radiocontrast to get an image of a digestive system was in 1908.

That radiocontrast was, and is, actually barium sulfate. It is a very dense liquid, heavy enough to use as a drilling fluid in oil and gas wells, and its opacity to x-rays paired with its low toxicity makes it ideal for gut mapping.

Natural barium always occurs as a salt. The metal alone has very few uses. It explodes when exposed to moisture. Two of its compounds make different shades of green fireworks. It is used in optics and electronics and YBCO superconductors.

The water soluble salt is highly toxic and used as rat and insect poison. It can cause anxiety, paralysis, blindness, and cardiac problems. The good news is it is not carcinogenic and does not accumulate in the body, so at least it won't give you cancer.

56

Wayfinder

With Bran and Cafall at his heels, Will stepped forward into a faint glow of light that caught him into such wonder that he could only stand and gaze.

-Susan Cooper, *The Grey King*

Barium is chemically quite reactive, so most of its history and its uses are all about combinations. Which is fitting; for what is the point of being a wayfinder if there's no one for whom to find a way? Its name, in Greek, means "heavy." But its compounds are what is heavy, not the metal itself. Barium's most essential cohorts are sulfur and oxygen, which together form a slurry that is so dense, it is used (like cesium) to guide rock chips out of mines during drilling; going into the darkness of the earth, leading the way out again. In addition, it is both fairly nontoxic and is opaque to x-rays. Hence a barium solution, introduced from above or below, is used to map digestive systems; again, going into the gloom, and leading the way. Shining a light on any problems it finds there.

A small square of mostly pure barium was always included in old vacuum tubes, because any stray gases that may sneak in would be gathered up like lost souls and kept from doing harm. Nowadays, barium has a much more exciting use, as a central character in YBCO superconductors, which defy conventional laws and could soon teach science much it didn't know. Barium, as usual, will stride forward into the dark, and find the way.

There is darkness ahead of you, according to the cards; now more than usual. And in that darkness there may dwell things that can hurt you. There are unknown monsters; there are pitfalls and polyps. Or are there? All you know is that there opens before you a pitch-black tunnel, which could in fact be smooth and safe but you will not know until you walk boldly into it, carrying light. And if it branches, what then? Which path will you take? A wayfinder is important right now. Is it who you are, or is it who you need? Can you do this for yourself, or do you need a posse? That may well be your best bet. Barium never travels alone.

Lanthanum

Element Number:	57
Weight:	138.90547u
Normal State in Nature:	solid
Melting Point:	1191° Kelvin
Boiling Point:	3737° Kelvin

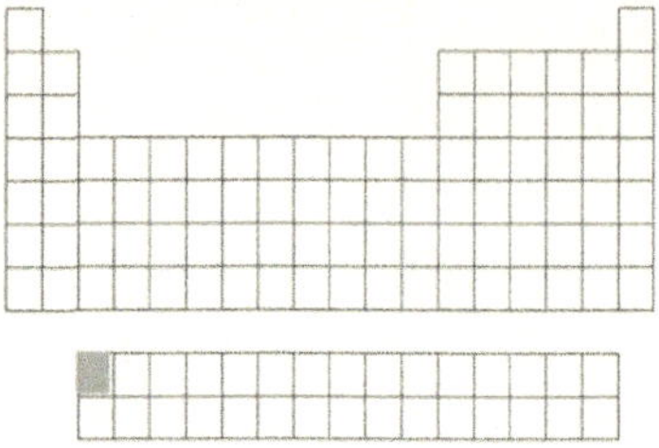

In the 18th century it often happened that someone would notice an undocumented mineral somewhere, send it to a chemist, and learn that they had been instrumental in finding a new element. Wilhelm Hisinger tried this in 1781, when he was just fifteen, but Scheele sadly did not detect anything unusual about the rock in question.

Hisinger grew up to be friends with Berzelius, and in 1803 they analyzed the mineral together and discovered an oxide they named *ceria*, after the recently discovered dwarf planet Ceres. At the same time, Klaproth found the same oxide, confirming their find. By 1843 ceria was revealed by Mosander to be a mixture of several new elements, which were damn difficult to isolate from each other.

In recognition of this, Mosander called this particular one lanthanum, "to lie hidden."

The lanthanides are fifteen metals which tend to be found in the same ores and share many traits. Lanthanum is three times as abundant as lead, but even with today's technology it is extremely hard to separate from other F block elements, collectively called rare earths. Most of its uses are actually uses for a mixture of lanthanides; as long as it's about 50% lanthanum, that's usually good enough.

It was used in old lantern mantles, and is in new Prius batteries. It binds phosphate for people with kidney failure. It shows up in telescope lenses and water treatment. It sparks dramatically, and is utilized in filmmaking effects as well as lighter flints.

It is moderately toxic. Do not eat it.

57

Spark

You're only given a little spark of madness. You mustn't lose it.

-Robin Williams

Welcome to the lanthanide army! We are all basically interchangeable.

When you see a frog with a silver shield, it could easily do the job of most of its cohorts. Rare earth elements come out of the mine mixed up, and are frequently left together as "mischmetal" because why go to the huge expense of separating the soldiers of this frog's army?

Just because lanthanum is fiendishly difficult to extract from its little brigade, does not mean it has no uses. Mischmetal is used in camp lantern mantles, lighter flints, cathodes, electrodes, and to make spark effects in movies. The majority of lanthanum's uses boil down to the spark, it seems.

Lanthanides are all so similar because the number of electrons in their outer shells is the same across the board, while an inner shell has a different population in each. Thus, while the chemical properties are nearly identical, the more subtle qualities—like magnetism—are quite different. This is to remind us, no doubt, that no matter how hard we may try to blend in, there will always be a layer within that sets us apart. Lean into that, says lanthanum. It is that spark which makes you *you*, bright and inspiring.

You don't have to be separate to be rare. You don't have to be aloof to be a leader. You have the power to spark right now—spark change, spark enthusiasm, spark inspiration. Is it fireworks and special effects, to awe and entertain, that you are craving? Or is it the steady light of a lantern, the joy in bringing together a mischmetal of people who might otherwise have felt like outcasts? Or would you rather look inward, to your own soul's electron orbitals that have too long been neglected, and draw them out with joyful excitement? Whichever way you go, you are the captain and the champion of the unseen. Go spark something.

Cerium

Element Number:	58
Weight:	140.116u
Normal State in Nature:	solid
Melting Point:	1071° Kelvin
Boiling Point:	3715° Kelvin

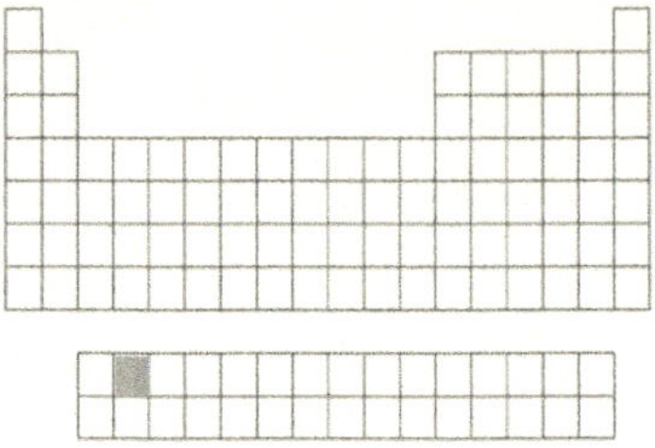

Hisinger was a mine owner and science enthusiast, who had an inkling that a particular rock found in and around his mine had a story to tell. He and Berzelius studied the ore that would eventually come to be known as cerite, and they got as far as an impure sample of an oxide they called *ceria*, named for the recently discovered dwarf planet that was named for the Roman goddess. Mosander, a cohort of Berzelius, took it a step further; and achieved much purer ceria, lanthana and didymia.

Oxides were as far as they could get. The rare earths are too electropositive to allow for a complete isolation until Davy invented electrochemistry.

Pure cerium was finally obtained in the 1830s.

Of course, its main applications actually make use of its oxide, so isolating it was not the breakthrough it can sometimes be.

Chemical-mechanical planarization, also known as CMP or, as the laymen call it, "polishing," is cerium's most important application. It also removes the green tint from glass caused by iron impurities. Cerium diodes emit white light. It's a large percentage of mischmetal, used as a flint in lighters. It can add photostability to pigments. It was key in making the cores for the atomic bombs used in World War II.

It may lower cholesterol and blood pressure, but if you eat it and it doesn't work, it won't poison you. Praise the goddess.

Abrasive

Narknon was washing Harry's face with her razored tongue, and Harry was trying not to mind.

- Robin McKinley, *The Blue Sword*

To be abrasive is not necessarily a bad thing. Cerium's highlighted trait is why it is used to polish glass in high-quality optics, which it does better than any other metal. It's also awfully common, especially for a "rare" earth; in mischmetal it is usually twice as plentiful as lanthanum, so it sparks like a good pyrophore should. And shows up to help in catalytic converters, and gives photostability to pigments and polymers. Why, then, is his sobriquet just "abrasive"?

There is a lot of fog in this world of ours; a lot of confusion about the way things *are* versus how they are *perceived*. One must scrub the constant and conflicting input, scour it rigorously until it comes clear. That is the objective of cerium, across the board: To polish, to bring clarity, to throw light—or at least sparks. And yes, when one applies these skills to their day-to-day life, there will be those who complain. You're too much, they might say. Stop checking facts, stop being so irritating. Stop correcting my worldview. Stop being so obnoxiously *abrasive*. One may even have this argument with oneself…and that is the time to have the brightest and best scouring of all.

Embrace the abrasive, my friend. Cerium is a sign that your lenses need polishing and your air needs clearing. Call people on their shit, especially yourself. Check all the facts and question the sources. Lies may be everywhere, but there is truth to be had, and there are those who care about sorting the real from the smog. Get out your psychic sandblaster and join them.

Praseodymium

Element Number:	59
Weight:	140.90765u
Normal State in Nature:	solid
Melting Point:	1204° Kelvin
Boiling Point:	3785° Kelvin

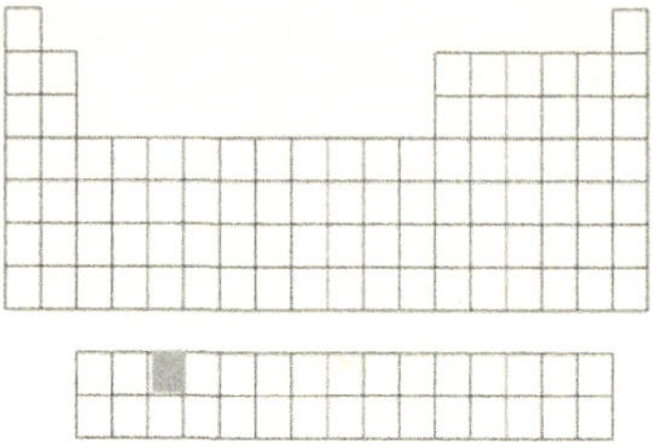

While Hisinger and Berzelius in 1803 were in Sweden, finding out things about the lanthanide-rich ore they later named cerite, Klaproth was doing the same thing in Germany. These simultaneous but independent discoveries were not an uncommon phenomenon.

None of them, however, managed to tease all the lanthanides apart. It is, to this day, a daunting enterprise. Though Berzelius' lab partner Mosander at least identified and named the oxides *lanthana* and *didymia*, he did not realize that didymia was two elements, not one, almost inextricably tangled.

In 1885 von Welsbach separated those two. He confirmed the separation with spectroscopic analysis, and named the new elements in Greek, as is traditional.

The *praseos* (leek green) part of the name comes from the color of its salts. *Didymos* means twin, possibly because of its tight relationship with its ore buddy, neodymium.

Praseodymium is isolated in order to be used as a yellow-orange ceramic stain. But mostly, as with its lanthanide brethren, it is not isolated before being used. Didymium makes lenses for glassblowers' and welders' goggles, which protects the eyes by blocking yellow light. It has other optical applications as well. It tints glass greenish-yellow, as well as making super stable magnets, used in wind turbines.

Praseodymium-doped silicate crystals can slow a light pulse.

Shade

You complain that your tree is not valuable as lumber. But you could make use of the shade it provides…

-Chuang-tse, *The Tao of Pooh*

Like the rest of the frog platoon, praseodymium is a rare earth jack-of-all-trades. It shows up in mischmetal and has skills in magnetic and electric applications. In fact, it can be swapped with most of its cohorts and make hardly a difference in functionality. But its interactions with light are what make it unique.

Praseodymium absorbs yellow light. Glassblowers, blacksmiths and welders use glasses with a special coating of 'didymium': A wondertwin mixture of neodymium and praseodymium. This makes searing white heat look like a soft glow, thus protecting the eyes of our daringest makers. Praseodymium is also used to stain ceramics and tint glass, giving varying shades of yellow, funnily enough, from orange-ish to greenish.

Shade is important, though often overlooked. Shade is a very specific form of protection that comes between us and the pain that too much light and heat can bring. When asked, What are the necessities of life? Most people would neglect shade, but anyone who has walked in a desert, or stepped barefoot on summer asphalt, or looked into a furnace will remember. Praseodymium doesn't provide general shelter, but it does bring on the shade, cool and inviting.

With praseodymium, the message is: Find shade. Your soul has been facing heat for too long, give yourself a break. Put something between yourself and the searing stress you face daily. Protect your eyes, especially the third one. Sunburn brings neither wisdom nor salvation, so stop courting it.

And finally, if none of this rings true; if you're thinking "I've done this self care already, what the heck"…Then your homework is, reach out to your most overheated, overstressed friend, and be shade for them.

Neodymium

Element Number:	60
Weight:	144.242u
Normal State in Nature:	solid
Melting Point:	1294° Kelvin
Boiling Point:	3347° Kelvin

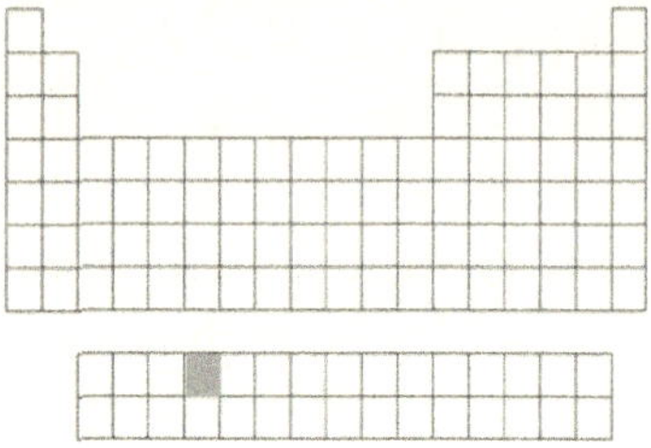

Found by Hisinger, pursued by Berzelius and discovered by Mosander is a common theme in the lanthanides. All lanthanides live in the same ores, and are very difficult to separate from each other. Mosander did his best to isolate one element in 1885, but he must have sensed that something was up because he named it *didymium*, meaning "twin." And it did, indeed, turn out to be twins: Praseodymium, and our favorite magnet material, neodymium.

Neodymium was isolated much more successfully in 1925, but the technology kept improving. In daylight, purer and newer neodymium glass is purple, but the older it gets, the redder it appears, because of impurities. The colors change dramatically with the ambient light, which is one of the reasons neodymium glass is so sought after.

But the biggest reason to go to the trouble of isolating neodymium is for magnets. Neodymium magnets can lift objects a thousand times their own mass, and can snap together hard enough to break bones. They have a myriad of applications, from headphones and hard disks to wind turbines and electric vehicle motors. There are undoubtedly several in your house, doing jobs you may not have even thought about. And as long as they don't get too hot or too cold, they will continue to do so.

Neodymium glass is not just for antique collectors; it's needed in high-powered telescopes. It's also used in lasers and has possible future applications as a substitiute for uranyl acetate when used as a contrasting agent for electron microscopy.

Attraction

Ask for what you want and be prepared to get it.

-Maya Angelou

Neodymium can be used as a glass colorant, giving us hues ranging from yellow through pinks and purples and all the way to blue, which change with the changing quality of the ambient light. They are very lovely, these unique color-shifting vases and bowls, and absolutely worth doing an image search for "neodymium glass" and OKAY FINE we'll talk about the magnets!

Once alloyed with some iron and a touch of boron, neodymium makes the strongest permanent magnets known to man. They are used everywhere, from the jewelry and music industries to electric car motors, and you certainly have some in your home. The pull of it can be frightening; even small ones can snap together from several inches apart, and if you happen to have a hand in the way, you will get bruised. Just as attraction is a law unto itself, which many of us have learned the hard way. Whatever we call into our lives, we must then bear the consequences.

There is another aspect to neodymium. Lasers require a "gain medium," which is a substance that can store energy and then release it when triggered, pushing much more out than was apparently put in. Some use liquids for this, some gases; and some use neodymium-doped glass or crystal, loaded up with photons. See? The glass isn't just pretty. And attraction isn't just for romance. It can reach out into the universe like a conduit of light, a beacon to bring home whatever we were missing.

With neodymium in your spread, now is a very good time to manifest. Call your heart's desire into being; imagine your best life so vividly it snaps into focus. Identify what you need and draw it to you, feel the joy of it before it even arrives. Make sure you've got the proper light quality to recognize it when it comes, and make space for it. Bruises may occur during the transition, but it may well be worth every one.

Promethium

Element Number:	61
Weight:	145u
Normal State in Nature:	solid
Melting Point:	1315° Kelvin
Boiling Point:	3273° Kelvin

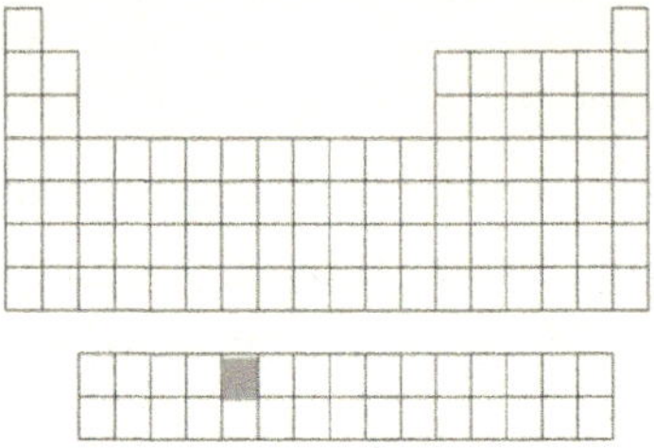

In 1902 it was believed that neodymium and samarium were consecutive lanthanides, but Brauner was unconvinced. He pointed out that their traits are too different, and proposed that there was an intermediate, undiscovered lanthanide; and chemists were off to the races. Rolla and Fernandez of Florence declared element 61 found in 1924, and named it florentium. Hopkins and Yntema of Illinois, in 1926, claimed the discovery, and called it Illinium. They were all wrong. A clue to the nature of elusive 61 came in 1934 when Mattauch published the isobar rule. In short, 61 *had* to be radioactive. So finally, in 1945, a team at Clinton Labs (later Oak Ridge National Laboratory) isolated 61 from uranium fission fuel products, the name suggested by the wife of one of the team members was assigned, and promethium was born.

"Both the daring and possible misuse of mankind's intellect" was the symbolism behind the name. The most stable isotope is promethium-145, with a half-life of 17.7 years. Most of its applications are in research and pure academics.

Promethium-147 has a few additional uses, such as atomic batteries, and paired with a phosphor to make glowing signal lights. None of these are widespread, as only a few hundred grams of promethium are produced per year, at the most.

Russia seems to use more of it than any other country. Make of that what you will.

Brevity

Well, maybe it's because I can recite all fifty states in a quarter of a second: YAP.
-Peter Griffin, *Family Guy*

Promethium is the only lanthanide that has no stable isotopes. It is unhealthy for living things to interact with, but the odds of that happening are slim. Promethium is the absolute rarest rare earth. And even where it does exist, it only does so briefly.

It took a very long time to identify, but then a very short time to fall out of favor for practical uses. It replaced radium in luminous watch dials for a bit in the 1960s, but was itself replaced by a hydrogen isotope. It can still be found in atomic batteries powering satellites and in submarine communications arrays; but any use made of it must take into account that it won't last forever. Its longest-lived (and most utilized) isotope lasts 17 years; the shortest, only a few nanoseconds.

One can accomplish a lot in seventeen years. Would it be easier, or harder, to do so if we knew that, afterwards, we would change so completely we wouldn't be the same person anymore? Not dead, just…different. And suppose that time were only four years? Or two? A week?... Depending on the isotope, that is how promethium exists. With a clock inside, counting down.

Promethium was named for Prometheus, as a reminder of "the daring and possible misuse of mankind's intellect." Both must be kept in focus, if these brief years are to be used well. Whether on land, in the sea or among the stars.

Do what is important right now. Pick up a project that excites you. Call your mother. Build a pillow fort with your kids. Read the book you promised yourself you'd read. This is not a message about mortality, but about radioactive decay. In a few years, believe me, you will be a different person, with different hopes and ambitions, different people to spend time with. Let that person have pleasant memories of this time, of this day. Give your older self some good stories to tell.

Samarium

Element Number:	62
Weight:	150.36u
Normal State in Nature:	solid
Melting Point:	1347° Kelvin
Boiling Point:	2067° Kelvin

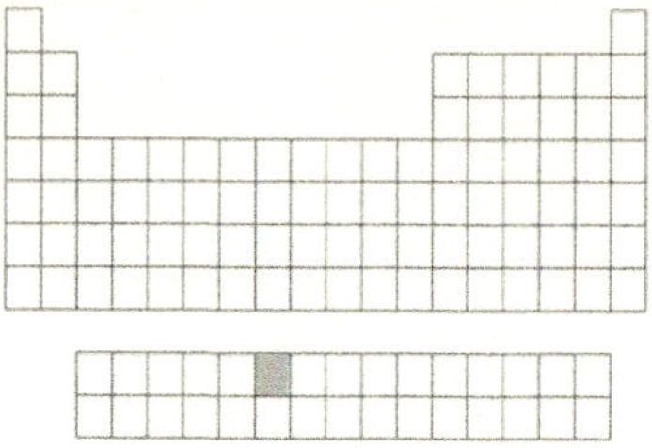

The competition to isolate all the lanthanides was fierce. Samarium was dibbed by several people at once, but most give credit to de Boisbaudran in 1879, who named it for the mineral he'd found it in, samarskite. Samarskite came from the Ural mountains, and was named after the man who had graciously allowed minerologists access to the area; and so samarium is the first element to be named for a person, albeit indirectly.

Samarium was merely a component of mischmetal until the 1950s, when new techniques were invented for separating the lanthanides. It is still more common to buy and sell the oxide, called samaria, rather than the pure metal itself.

One praiseworthy use for samarium is as a chemical reagent, neutralizing environmental carcinogens and even helping to decompose plastics. One of its isotopes is used in nuclear reactor control rods, as it has a high cross-section of neutron capture, and another isotope (radioactive, this time) is utilized in cancer-fighting medicine.

But its most widespread application is samarium-cobalt magnets, which are nearly as strong as the neodymium kind and, as a bonus, stay magnetic at much higher temperatures. While neodymium will lose its magnetic properties at 400°c at most, samarium's curie point is 700°c. These magnets are used for headphones, small motors, and pickups for various electronic musical instruments.

Sanguine

You sanguine about the kind of reception we're apt to receive on an Alliance ship, Cap'n?"
"Absolutely. What's sanguine mean?"
"'Sanguine'. Hopeful. Plus, point of interest? it also means 'bloody'."
"Well, that pretty much covers all the options, don't it?"

-Firefly

Musicians are familiar with samarium-cobalt magnets in their guitar pickups, which are slightly less powerful than neodymium but stay magnetic at much higher temperatures. This may not be practical, since the temperature during a performance is unlikely to approach 700ºc, but perhaps it is superstitious. Musicians have to function well when the heat is on, so knowing their guitar pickups have that property could be a morale boost. Beyond the magnets, one of samarium's isotopes is the basis for a radioactive medicine that targets mutated cells, and is used to treat a wide range of cancers. Samarium is also an important chemical reagent, used in the separation of pollutants. And, finally, it is a crucial performer in nuclear power control rods, absorbing neutrons beautifully. A sanguine temperament will always save the day.

Medieval doctors believed that the human body was ruled by four "humors," each with a physical aspect and a metaphysical affect. "Sanguinity" lived in the redness of blood, they thought, and provided a person with the qualities of air: confidence, optimism, vitality, and hope. If it has a fault, it is that too much will make a person reckless, egotistical or overly dramatic. As, indeed, it is a bit dramatic to use an extremely heat-tolerant magnet where extreme temperatures will never be reached. But, oh well. Samarium is still, as they say, bloody brilliant.

By drawing samarium, you have drawn on the power of sanguinity. It is time to welcome feelings of optimism and hope. Good things will happen, and you can help them. Whether your efforts are aimed at healing people, restoring the environment, or preventing disaster, you got this. Trust yourself to stay strong when the heat is on, and welcome the morrow with music.

Europium

Element Number:	63
Weight:	151.964u
Normal State in Nature:	solid
Melting Point:	1095° Kelvin
Boiling Point:	1800° Kelvin

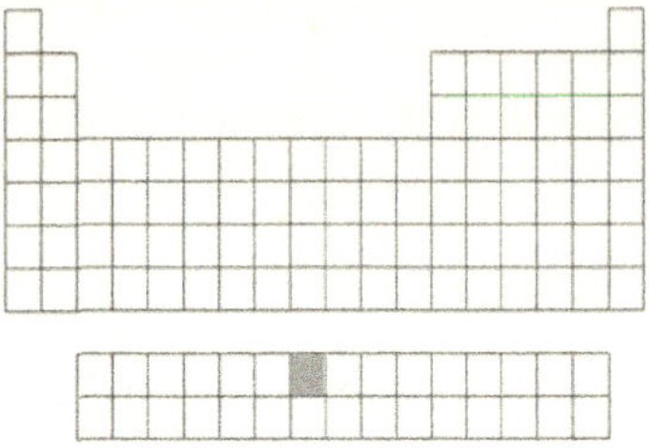

Europium, though it had been observed in the late 1800s and officially isolated by Demarcay in 1901, was not initially one of the popular lanthanides. Spedding invented the ion exchange method of separating the lanthanides from each other in the 1950s, and thereby revolutionized the industries that make use of rare earths. It was then found that europium (named for Europe, of course) is the softest, most reactive, and one of the rarest, and though many of its compounds are phosphorescent, that did not make it useful. For a while, europium was a mere curiosity. However, when in the 1960s it turned out be a necessary component of yttrium orthovanadate, a new, brilliant red phosphor that would completely change color television, there was a scramble to buy up the hitherto interesting but useless metal.

Europium-based phosphors can be red or blue, and have a myriad of niche uses. Lasers, monitors, fluorescent lights, as well as anti-counterfeiting Euro banknotes. Even glow-in-the-dark toys, which typically use strontium-aluminate phosphor, are doped with europium. Very little is needed to achieve the desired effect, which is fortunate because very little is available.

There may be a future in which europium is needed for something other than things that glow, and in that case we must hope that deposits of it turn up somewhere. Even lanthanide-rich minerals only have, at absolute most, 2% europium. Perhaps it's good that its uses are not terribly demanding.

Luminosity

So I decree that if she wakes all night she shall wax and wane with its mistress, the moon.

-George MacDonald, *Little Daylight*

Europium is an exceedingly rare rare earth, and some of its uses are rather quaint and old-fashioned. It was a critical component of cathode ray tube televisions, because it provided the bright red phosphor dots within the screen. It is also one of the elements that gives those swoopy compact fluorescent lightbulbs a pleasant daylike spectrum. Of course, CRTs and CFLs are falling by the wayside with the proliferation of LED technology, and for a long time the demand for europium waned.

But like the Moon, europium is waxing again. Its compounds make the strongest (and most expensive) glow-in-the-dark paints; it can be utilized as a doping agent in lasers; and some clever people are discovering that it may even be used in LED screens themselves to make them exponentially more effective. And quantum memory chips may sound like fiction, but the people inventing them as we speak are using europium to do so.

The Moon glows, and it waxes, and it wanes. So do we, as we notice when we're paying attention. Our energy lags, then increases; our interests fall by the wayside, and new ones emerge. Europium exemplifies this pattern, and most importantly, through it all, it is luminous.

Now is a fine time to begin to observe your own patterns. Human beings are as apt to wax and wane as the Moon, but we are a little less reliable and certainly not uniform. So watch your energy levels, and notice: When is your light strongest? When is it weakest? What is your most creative time of each day or week or year? When do you feel as if you're a wan sliver of a crescent? Whatever your rhythms are, and whatever you do with this information, never forget that, soft or bright, you shine. You are wondrous. You are a luminous being.

Gadolinium

Element Number:	64
Weight:	157.25u
Normal State in Nature:	solid
Melting Point:	1585° Kelvin
Boiling Point:	3545° Kelvin

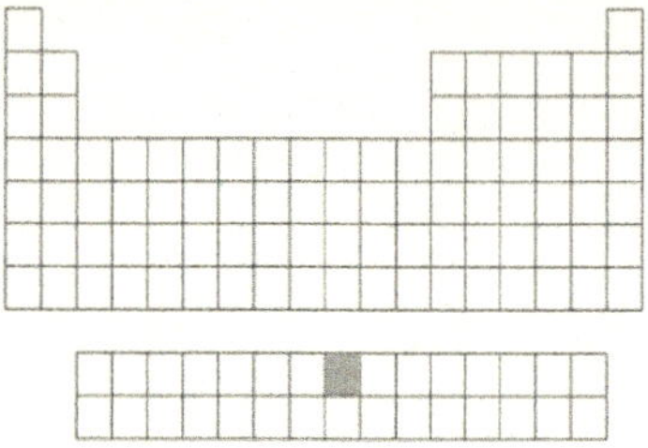

In 1880, de Marignac recognized that two minerals, cerite and gadolinite, were displaying a spectroscopic line that did not belong to any known element. Since there was clearly more of the mystery substance in cerite, that is the rock he investigated until he had isolated the oxide of a new rare earth metal; he named it gadolinia, after the *other* mineral, which had in turn been named for Gadolin, the Finnish geologist. It is possible that de Marignac chose this name because ceria had been taken already, 78 years earlier.

It was de Boisbaudran who isolated the metal itself, in 1886, but de Marignac gets credit for the initial discovery, and the name.

Gadolinium's uses are all over the map; if said map is a map of nuclear medicine, and nothing else. Gadolinium is used as several unique contrast agents in MRIs; a slightly different compound is a phosphor used in x-rays; and it is particularly useful in detecting brain tumors, as the contrast will only penetrate a compromised blood-brain barrier. In addition to detecting them, radioactive gadolinium-based medicines have been used to target tumors.

It does have a smattering of other uses, from reactor shielding to optics, but nuclear medicine is its favorite place.

Infiltrate

Feast your eyes on its total inability to be seen!

-Tom Servo, *MST3K*

Gadolinium is one of our specialist frogs. While it's not as generally useful as the other rare earths, it has the distinction of being crucial in medical imaging. MRIs would be useless without gadolinium contrast, which is injected into a patient so their bloodstream can be observed. It infiltrates every capillary, and like a spy sends back critical information.

Another skill of this ninja element is in neutron radiography, which is also a technology of collecting data on that which is difficult to see, this time in aerospace engineering. A gadolinium "neutron conversion screen" is aimed at a turbine or engine, where it can absorb neutrons and then emit high-energy electrons as pictures that look rather like x-rays. This is used to develop machine technology, and for diagnostics when things aren't working. Gadolinium's skills at espionage are not limited to medicine.

It has a further trait that sets it apart: gadolinium has an unusual curie point. This is the temperature beyond which a magnetic material ceases to be magnetic. For most ferromagnetic metals, that temperature is in the 500ºC range, but for gadolinium, it's room temperature. It can change before our eyes from a cool magnet to a warm lump of inert metal. A master of disguise, as a good spy should be.

With gadolinium in your spread, it may be time to get sneaky. Gather some information. Be subtle about it, but thorough. Whether facts are being deliberately withheld from you, or whether circumstances have simply been opaque to your perception, open your eyes and gather some knowledge. Unleash your inner ninja spy, and infiltrate.

Terbium

Element Number:	65
Weight:	158.92535u
Normal State in Nature:	solid
Melting Point:	1629° Kelvin
Boiling Point:	3500° Kelvin

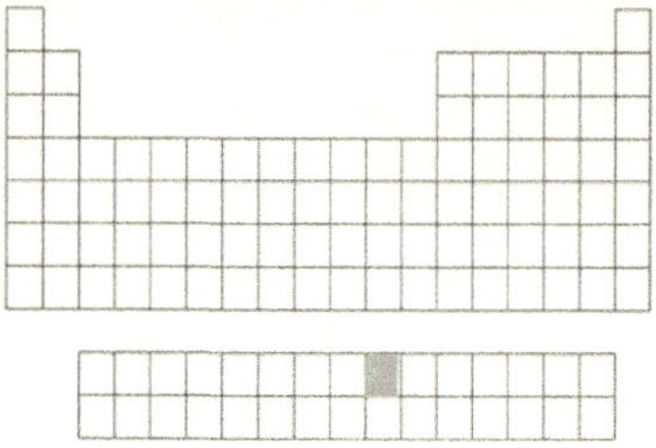

In the 1840s, Mosander decided that impure yttrium oxide warranted further study. He broke the sample down into three distinct substances, which he called yttria, terbia and erbia; all oxides. Terbia had a pink tint to it, while erbia was tinged brown but essentially colorless.

Other scientists looking at his work argued that this so-called "erbia" possibly did not even exist, and during the debates the names got permanently swapped. What Mosander had first called erbia became terbia, the oxide of a rare earth metal that certainly does exist, even if its oxide is brown, not pink. It also fluoresces bright yellowish green under UV light, but in 1843 they did not know that.

Its oxide is a phosphor used in all trichromatic lighting, providing the lemon yellow color. This phosphor is also used in artificial daylight lightbulbs, so a modern 13 watt bulb makes as much light as an old 75 watt one did, and your eyes can't tell the difference. Like most lanthanides, it's a big player in solid-state electronics.

Magnetostriction, in addition to being a really cool word, is the ability for certain substances to expand and contract measurably when under the influence of a magnetic field. This is a quality terbium has, giving it applications in actuators, buttons, sonar, sensors, speakers, and other magnetomechanical things.

Amplify

And I played piano the same way, seeking salvation through volume.
-Amanda Palmer, *The Art of Asking*

Terbium changes its shape in the presence of a magnetic field. Not dramatically, like some kind of TX1000, but enough that if you measure a rod of terbium near a magnet, and then take the magnet away and measure it again, you will have two different numbers. This gives terbium a variety of uses, in sonar and sensors and buttons; it also gives it the power to turn any flat surface into a speaker, which is pretty darn cool. The ability to amplify a signal—to turn something barely real into an audible communication—is very nearly a superpower.

Terbium is also the main component in a certain green phosphor, which is itself useful across the board. Biochemists use it, electronics utilize it, and phosphor-based trichromatic lights would be useless without it. Its output of light far exceeds what an incandescent rig would do with the same amount of electricity; thus does terbium double down on its role of amplification.

When we amplify our own voice, we do run the risk of being told to shut up. When we phosphoresce, we may well be told we're tacky or attention-seeking. But terbium is way too cool to give a darn, and perhaps therein lies the key message. Being heard is a good thing.

Sometimes it is hard to boost your own signal. Sometimes you get lost in the white noise of your own mind and can neither hear nor be heard by other humans. And when you are feeling unheard, you almost always feel unseen. This could be one of those times.

So drawing terbium right now is a clarion call: Just give yourself a charge, and you will shine brightly. Find a large surface, and you will find your voice. Get as loud as you need to, and for good measure, go ahead and glow.

Dysprosium

Element Number:	66
Weight:	162.5u
Normal State in Nature:	solid
Melting Point:	1685° Kelvin
Boiling Point:	2840° Kelvin

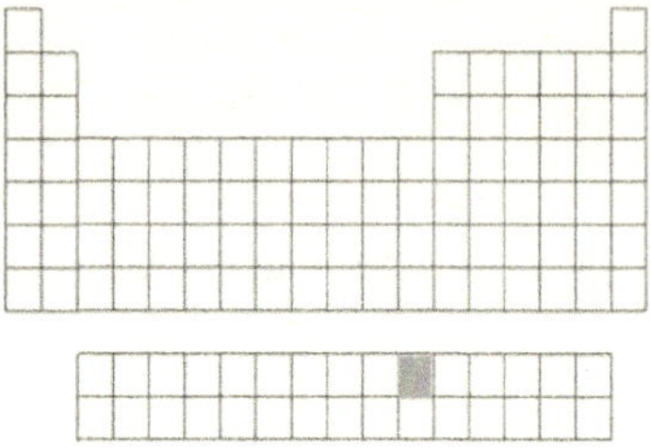

Holmium oxide was found by de Boisbaudran to contain a second oxide, which he extracted and then tried to purify. This turned out to be not as simple as it sounds. It took de Boisbaudran more than thirty tries to separate this hideously stubborn rare earth from its oxide.

When he finally succeeded, he commemorated the process by naming this new element *dysprosium*, which is Greek for "hard to get." This triumphant discovery happened in 1886, but it was not until the 1950s that ion exchange techniques made it possible to consistently produce pure, untangled lanthanides.

It is used in control rods, in lighting and lasers; in lamps, and magnets, and fuel injection devices. Dysprosium can pretty much do the job of any other lanthanide. The flip side of this is that pretty much any lanthanide can do every job dysprosium can do. Its uses are widespread, but incidental; and do not usually require that it be isolated before being used. It is even one of the components of the terbium alloy that is highly magnetostrictive, and so is included in all of those applications as well.

It is useful, just not terribly singular.

Ancillary

Probably just some local fisherman, out for a pleasure cruise, at night…in…eel-infested waters…

-Vizzini, *The Princess Bride*

In all the places where rare earths are used, you can find dysprosium. In magnets and halide lamps and lasers and even nuclear control rods, a small smattering of dysprosium is probably present. But what does it do? Well, at very low temperatures it is highly magnetic; and…that's it. Beyond that, it is there. It exists.

Clean energy takes advantage of dysprosium's magnetic properties and includes it in wind turbines and electric car motors. It does add longevity to permanent magnets, and it even luminesces in the presence of ionizing radiation, but who doesn't? When you look at dysprosium's uses, it's all just more ways dysprosium is very like the rest of its froggy brethren.

Sometimes we are simply here. Sometimes our grand plans leave us at a loose end and we get to kick our heels for a while. We can be useful, sure. We still have inherent value and worthwhile skillsets. But we don't need to be the main attraction. We are a small component in a big world, but there is no shame in that. Dysprosium does a lot of neat things without ever being the biggest frog in the pond.

When you draw dysprosium, you have the opportunity to just hang out for a moment. You can make yourself useful if you like; your skills and value are still prized, and you do bring extra oomph to the party. You would be missed if you weren't there, but things will go fine if you don't interfere. The pressure is off. Your showtime is not now. Sit back, relax. Observe. Help if it's needed. And if you wish to engage your own personal magnetism, the best way to do so is to keep your cool.

Holmium

Element Number:	67
Weight:	164.93032u
Normal State in Nature:	solid
Melting Point:	1747° Kelvin
Boiling Point:	2968° Kelvin

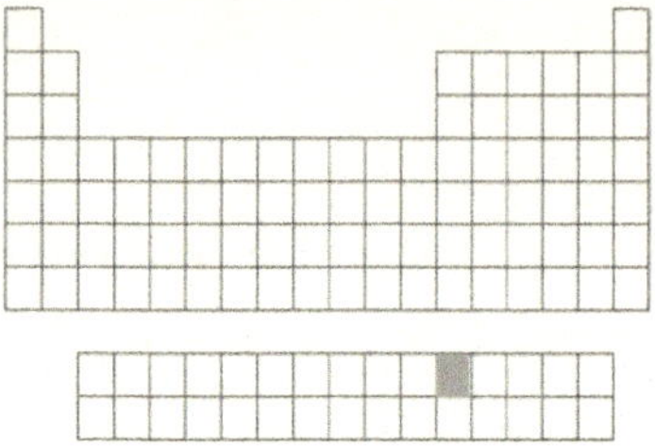

In 1878, Cleve used Mosander's methods to isolate two new oxides from a sample of erbium. He called the brown one *holmia*, after the Latin name for his native Stockholm; and the green one *thulia*, after the Greek name for Iceland.

Incidentally, at the same time, holmium was being discovered by Soret and Delafontaine, who spotted it by its spectrographic absorption bands. They rather dramatically called it Element X, but Cleve isolated the oxide first and so his name for it was kept.

As with every other lanthanide, it was not actually purified until the 1950s, when the ion exchange method of separation was invented.

Holmium is the strongest magnetic material known; when placed in a preexisting magnetic field, it concentrates and amplifies it. Holmium pole pieces, or "magnetic flux concentrators," are integral to MRI machines and other scary-high strength magnets. It is also used in lasers to do essentially the same job of making them stronger; holmium-doped YIG lasers have medical and dental applications.

Holmium is also used as a yellow or red colorant for cubic zirconia; as well as glass, which is then used to calibrate spectrometers. It is a neutron poison in nuclear reactors.

It is being studied for use in quantum computers for data storage. Stay tuned for that one.

67

Focus

Treefallinginthewoodstreefallinginthewoodstreefallinginthewoods…

-Bart Simpson

Holmium is a rare earth; exceptionally rare, in fact. It is rather difficult to tease it out of the jumble of rare earths contained in certain minerals; but we do, because we need holmium. It makes our most important magnets exceptionally stronger.

Lumps of holmium are placed on the poles of the magnet in MRI machines, which causes said magnet to become frighteningly effective. This is because holmium pole pieces concentrate magnetic fields. One does not find plain holmium magnets; its forte lies in focusing a field that already exists, in drawing together the lines of force and giving them more power through proximity to, and alignment with, each other.

It has a similar job in YAG medical lasers; holmium takes what is already there and gives it more oomph by storing optical energy, then releasing it. Like a desert rain frog practicing tai chi, holmium knows how to move energy around, store it, release it and, most of all, focus it.

According to the cards, now is a very good time to focus on the flow of your energy. If you have been feeling like you're all over the place, that your energy is being scattered to the winds so that you can meet the expectations of others; if you are feeling like life is a whack-a-mole game, now is a good time to pause. Breathe. Gather your energy; direct it where you want it to go. Also (though you may hate this advice) take 30 minutes once a day and do something physical, like a walk or yoga or tai chi. Your body is the home of your magnetic field, after all. If you want a strong ability to focus, align that first.

Erbium

Element Number:	68
Weight:	167.259u
Normal State in Nature:	solid
Melting Point:	1802° Kelvin
Boiling Point:	3140° Kelvin

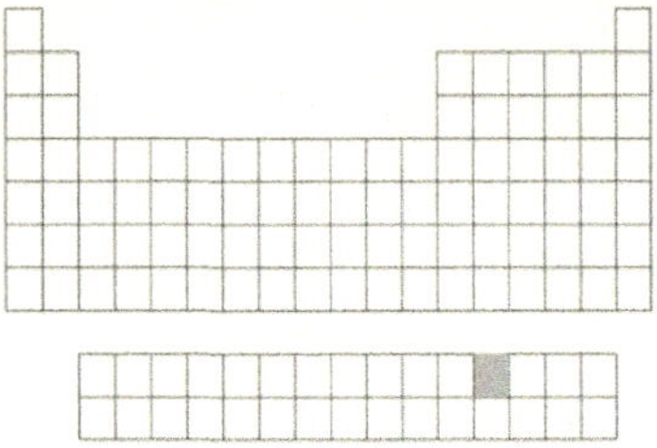

Mosander was working with what he thought was the oxide ytterbia, isolated from some gadolinite. He realized it was impure, and from it he isolated two other oxides, which he called *terbia* and *erbia*. Even these were not pure, lanthanides being what they are, and scientists began to argue about the samples' veracity.

During the kerfuffle, some random spectroscopist switched the names of the two oxides. Though discovered in 1843, the names were accidentally but permanently swapped; terbium became erbium in 1860, and so it remains.

Pure-ish erbium metal was not produces until 1934, and it was not until the 1990s that it became affordable enough to be used as a glass colorant.

Erbium-doped glass is used in lasers and in optics, doing serious things in surgery and communications. But since erbium brings a lovely shade of pink, it is also used in art glass, porcelain and cubic zirconia.

It is sometimes used in metallurgy, because it makes alloys more workable. It is also sometimes used in neutron-absorbing control rods in nuclear reactors. And it shares many traits, and therefore uses, with the rest of the lanthanides. It's a component in mischmetal and a gain medium in fiber optics.

Magnify

The world is so much larger than I thought. I thought we went along paths--but it seems there are no paths. The going itself is the path.

-C.S. Lewis, *Perelandra*

Erbium and Ytterbium are twins of function, if not in form. So, two pages from now there will be waiting a rather familiar description. Hang in there.

Erbium is, naturally, a rare earth metal, used to "dope" glass rods so that the pulse of light going in is magnified coming out. Erbium is not the first or the last element that can store energy and then use it to make a bright light; the uniqueness lies in the details. Erbium is hands-down the best for, specifically, fiber optics. It can take a weak pulse of illumination and use its own stored-up excited electrons to give that light more intensity *without* scattering it. This is important for optical communications. Lacking direction, amplification would be pointless.

This enhancement of a pulse can be done in other ways, too, but those tend to involve converting the light into an electrical signal and back again. Erbium requires no such translation. It simply gives the light the optimal wavelength for travel, and sends it on its way. To gather trustworthy input and send it out again clearer, sharper, larger and brighter…That is the gift of erbium.

Erbium in your spread is a "Boost the Signal" message. Communicate clearly, and make your output large and bright and, if possible, visual. The teaching aspects of your life would benefit from having your focus for a bit; so whether you are officially a teacher/trainer/coach or just a human with wisdom to pass along, do it. Do it large. Also, remember that erbium doesn't need to change the light in order to magnify it; in the same vein, try not to censor your message. Send it out as it came in, straight and true and magnified.

Thulium

Element Number:	69
Weight:	168.93421u
Normal State in Nature:	solid
Melting Point:	1818° Kelvin
Boiling Point:	2223° Kelvin

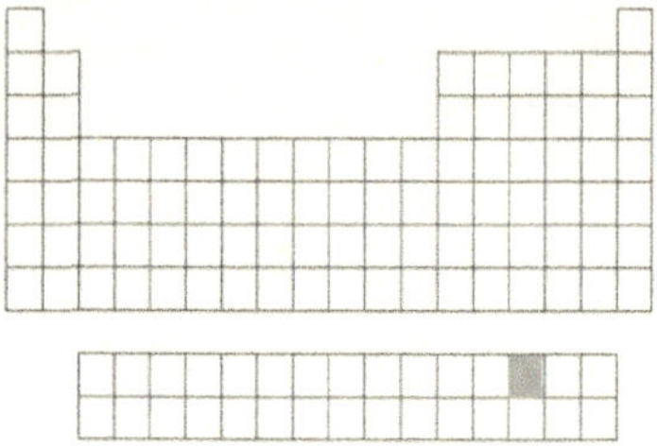

Cleve followed Mosander's methods in 1878, and isolated two new oxides from an erbium sample. He called the brown one *holmia*, after the Latin name for his native Stockholm; and the green one *thulia*, after the ancient Greek name for Iceland or Scandinavia.

Like all lanthanides, thulium was nearly impossible to isolate. Not only was it completely tangled up with adjacent elements, but it was so rare that, if you purified it, there would be none left. Still, 19th century chemists chased it, following its distinctive green spectral signature. James was the first to have any success, in 1911. The ion exchange separation method invented in the 1950s made thulium commercially available, though still ruinously expensive.

Thulium is used in arc lighting, for its green emission spectrum line that no other element has. It is also used for other common lanthanide-heavy applications; as a dopant for lasers, as a phosphorescent ink on banknotes, and (when irradiated) as a radiation source for x-ray machines. It has shown promise in high-temperature superconductors. It pals around with the other lanthanides.

It is still expensive as heck.

Specificity

One thing. Just one thing. You stick to that and the rest don't mean shit.

-Curly, *City Slickers*

Thulium is a bit of a nobody in the lanthanide army. It's the second rarest rare earth and hard to isolate, so for a long time it was just left as a supporting character in mischmetal and that was that. Even when a process was invented that could separate the rare earths from each other, thulium continued to be pretty interchangeable with all the others. Sure, it can be used in YAG lasers; and its synthesized radioactive isotope is great in portable x-ray machines. But so the heck what? Those things aren't specific to thulium.

And then there's arc lights.

Thulium's emission spectrum signature contains lines of green found nowhere else. Though the rest of the world may feel that thulium is a ditto mark on the periodic table, lighting designers know it by name and love it to pieces. It provides something essential to a small and select facet of the human race, the people who literally bring light to the rest of us. Who could ask for a better recommendation?

We all have some unique gift within us, though it may take time to find it. We are all indispensable to someone, though we may not be aware of it. One of our jobs in this world is to find our gift, and it might take a while. But even before we discover it, it makes the world brighter.

Thulium wants you to realize how important you are. You may be feeling as though you are surplus to requirements, just a cog in a machine or a face in a crowd. But this frog is here to tell you: You are something special. You have a spark no one else has, and it is vital. You are loved, and you are worthy of love. Ask yourself: What is the best thing about me? And from the depths of your soul, an answer will come. This is the part of you that you should be focusing on, honing and strengthening. This is the thing that is needed. This is your green spectral line.

Ytterbium

Element Number:	70
Weight:	173.04u
Normal State in Nature:	solid
Melting Point:	1092° Kelvin
Boiling Point:	1469° Kelvin

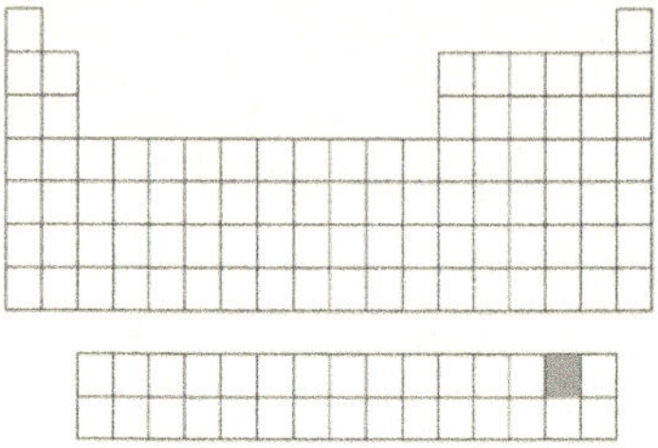

De Marignac, in 1878, took a sample of impure erbium and broke it down further. He discovered a new "earth," which he named *ytterbia*, believing it to be the oxide of a new element. In 1907, it turned out not to be quite that simple, when Urbain separated ytterbia into two distinct oxides which he called neoytterbia and lutecia. Welsbach, over in Austria, was doing the same experiments at the same time (as was James, who did not press his claim). Had Welsbach been luckier, these two elements would be known as aldebaranium and cassiopeium. But Urbain was on the international committee which decided who got priority of discovery, and though we don't want to accuse him of anything unsavory, the fact that these elements are now called lutetium and ytterbium (he dropped the *neo*) tells its own tale.

An up-and-coming use for ytterbium is in freakishly precise atomic clocks, which, if application catches up with theory and experiment, would be accurate to within a second per 13.4 billion years.

Other than that, its uses tend to be centered around solid-state lasers. It is a doping medium, which, in the right amount, will amplify the power of the beam considerably. And, of course, it can do all the lanthanide things; it improves the grain of stainless steel, can be irradiated to use in portable x-rays, and shows up in mischmetal.

Another possible use in ytterbium's future is in quantum computers. Cool.

Intensify

Our work is never over.

-Daftpunk, *Harder Better Faster Stronger*

Like erbium, ytterbium makes coherent light possible. In this case the technology that relies on the element is the solid-state laser; just a tiny concentration of ytterbium in the active laser medium will amplify the beam of light, making it intensely bright while maintaining the ordered flow of photons. Ytterbium lasers are long-lasting and dependable, and could even be the key to "power scaling," a concept that will add to the breadth of the effectiveness of lasers in general.

There are a lot of lanthanides that do basically the same job: Take input, make it bigger. Or brighter. Or louder; or just generally *more.* Ytterbium's ability to intensify a light beam is the reason why it is the heart of some lasers; and in science and industry and medicine, lasers are a crucial technology. Even if we have barely scratched the surface of what they are capable of. Even if there is a lot more work to do.

Ytterbium can also do a fair number of other things: improving the grain of stainless steel, for instance. Changing the hue of arc lighting. Stuff like that. But lasers need ytterbium's ability to turn up intensity; a skill we all have, and possibly need to be reminded of, from time to time. (Laser laser laser.)

*With ytterbium in your spread, the time has come to turn it up. Whatever you are doing, do it more. Whatever you are feeling, strive to feel it intensely. Wherever you are going, stay laser focused on it (*laser*), and do not let anything distract you. Are you happy? Dance with joy. Are you grieving? Tear apart a pillow and wail. Are you bored? Wallow in ennui and go get a tattoo. There is no room here for coasting, no autopilot, no waffling. Point yourself at the universe and flip the On switch. Go.*

Lutetium

Element Number:	71
Weight:	174.967u
Normal State in Nature:	solid
Melting Point:	1936° Kelvin
Boiling Point:	3668° Kelvin

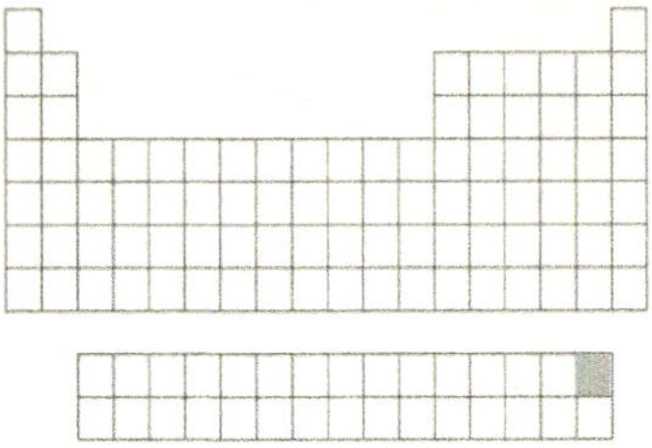

Urbain and Welsbach and James, all three at once, in 1907, discovered both ytterbium and lutetium. Welsbach and Urbain fell to arguing precedent, but Urbain had published first (barely) and so was given priority by the international committee (which he was on) in 1909. When it transpired that Welsbach's sample of lutetium had been pure, while Urbain's had not, many German-speaking scientists protested by calling element 71 *cassiopeium* in their papers. This persisted until the 1950s.

James, meanwhile, stayed out of the drama and merely stockpiled lutetium.

Lutetium, by the way, was named after the Latin word for Paris.

Even in lanthanide-rich minerals, lutetium accounts for only 0.0001% of the mass. It is extremely rare and expensive and difficult to isolate, and not really worth the trouble since its applications are echos of everything all the other lanthanides can do.

It can be a catalyst, a dopant, a lens material; it can be used in magnetic bubble memory and atomic clocks. It can help discover the age of meteors. LED light bulbs use it for color balancing, and is being explored for use in pain control for cancers.

The other lanthanides do most of these jobs as well for a lot less money.

71

Anonymity

When you are in a tree, be a tree.

-Jane Yolen, *Become a Warrior*

With all the mix'n'match going on amongst the rare earths, it is no surprise that lutetium's uses are shared by so many of his cohorts. In fact, the only thing that stands out about him is that he entirely fails *to* stand out. There is honestly nothing that he can do that another lanthanide can't do, probably better and definitely cheaper. He has trifling applications across the board, from nuclear medicine to arc lighting to circuitry, but none of these are unique to him. So he's pointless, right?

Of course not. Anonymity may not be glamorous, but it is a perfectly valid existence. As a minor player in so many arenas, and a lesser component of so much mischmetal, who are we to argue if the universe gives lutetium the right to abide in the background, unremarked but watchful?

Cast no aspersions on those adjacent to greatness. Lutetium may never have the pull of neodymium or the radiance of europium, but its place at the back end of the lanthanides gives it an amazing vantage point from which to view its fellows. Someday, one of the experimental uses for lutetium may prove to be its big breakout role, but until then, we should let it teach us how to blend in with the furniture and be anonymous.

It may seem like a strange message, but lutetium is a hint that today's big story is not about you. Your place in this situation is in the crowd, in the shadows or in the shrubbery. This is no hardship, but rather a reprieve; the eyes of Fate are elsewhere. You can take a breather. Pick your nose if you want to, nobody's looking. Grab some popcorn and watch the goings-on, you have an excellent view of them. Be ready to leap into action when the company moves.

Hafnium

Element Number:	72
Weight:	178.49u
Normal State in Nature:	solid
Melting Point:	2504° Kelvin
Boiling Point:	4875° Kelvin

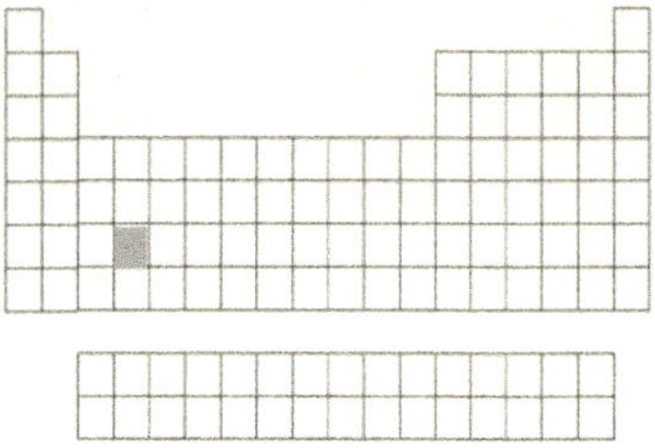

In the 1800s, the elements were organized according to their atomic weight. This was not always adequate, and some got misplaced. But when Moseley in 1914 used x-ray spectroscopy to discover a direct relationship between spectral line and precise nuclear charge, the gaps became evident and chemists began to scramble to find 72. Urbain insisted that he had already done so back in 1907, while studying rare earths, and had named it *celtium*. It was not easy to convince him otherwise. In the 1920s, Bury suggested (and Bohr agreed) that, being in the group with titanium and zirconium, 72 was most likely not a rare earth at all. Coster and Hevesy took this cue, and in 1923 they found element 72, and named it hafnium after the Latin word for Copenhagen, where their discovery took place. Van Arkel and de Boer isolated the metal itself in 1924.

It is fairly rare and difficult to isolate, but it still has a few uses. Control rods in nuclear reactors, for example. Its neutron capture cross-section is impressive, especially when compared to zirconium. Hafnium and zirconium are twins. Chemically you can't tell them apart, except that a piece of hafnium weighs twice what an identical piece of zirconium does.

Its ability to easily shed electrons makes it irreplaceable as the electrode tip on plasma cutters. It shows up in some electronics, some alloys and some catalysts; Apollo Lunar Modules contain it in their rocket nozzles, because it prevents protective oxides from flaking off the metal.

Dissever

And in the last instant the point of the spear was there before her, and it blinded her with light.
So she did not see, only heard, a kind of splintering sizzle. She had an impression of fireworks and bits of wood…

-Tanith Lee, *Black Unicorn*

Hafnium is always ready to facilitate a severing. It is used in control rods, because its capacity to cut off a nuclear reaction is impressive. It does this by absorbing neutrons. An interesting message, that: Can we sever a toxic connection through acceptance? Is that a thing?

Hafnium's only other significant application is in plasma cutters. The hafnium button embedded in the tip of a plasma cutter is heat-resistant and corrosion-resistant, reliable and long-lasting. It is crucial because it gives the plasma stream a heat boost, while letting the machine run cooler. All this is due to its ability to shed electrons at an unparalleled rate. So, severing a connection is about letting go then, is it? That certainly seems more intuitive.

But nothing is ever that simple. When we want to let go: of someone or something we have lost; of a toxic situation; even of our own dependencies or thought patterns…we can look to hafnium for our options. Should we cut off something hurtful by accepting it as it is, thereby cancelling out its control over us…or should we relax our grip on all the little things at once, slicing through the unwanted connection with a precise, cauterizing heat? The choice is ours.

With hafnium in your spread, you should look at letting go. What connection needs severing? Take the sword of hafnium and chop through anything that is holding you back, weighing you down or causing you strife. Choose well and be precise; hafnium could be about what's going on inside your head, rather than (or in addition to) your external circumstances. Remember that this severing can be done through radical acceptance, or through emphatic rejection. Choose which edge of the sword will serve you best, and swing that sucker.

Tantalum

Element Number:	73
Weight:	180.94788u
Normal State in Nature:	solid
Melting Point:	3293° Kelvin
Boiling Point:	5730° Kelvin

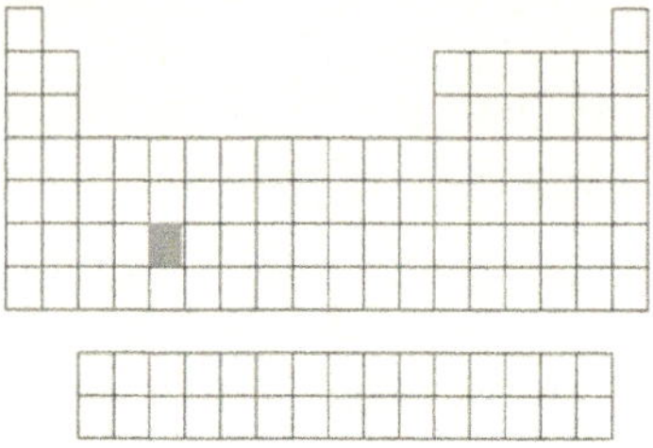

In 1801, Hatchett discovered columbium. In 1802, Ekeberg discovered a new, related metal, and named it *tantalum* because of its insolubility in acid. But in 1809, Wollaston examined the two oxides and declared them identical (despite having significantly different densities), threw out the name columbium and kept tantalum. Wöhler confirmed this. And so things stood until 1846, when Rose opened the debate again by insisting that this oxide was neither one nor two but *three* separate elements, tantalum plus two he called niobium and pelopium, after Tantalus' children. He turned out to be sort of right; pelopium was a mixture of tantalum and what had been columbium, now resurrected and renamed niobium. Several chemists confirmed tantalum's existence, and an impure sample of the metal was isolated in 1864 by de Marignac.

A vast number of electric devices require tantalum. As a material for extra tiny capacitors and resistors, it is unmatched. It is also considered a "refractory metal," that is, a metal with an extremely high melting point, which is both strong and ductile. It is made use of in various alloys and superalloys, in all sorts of places from rockets to nuclear reactors. It resists corrosion and is biologically inert, so can be used in orthopedic implants.

Sadly, its desirability has made it a conflict resource, linked to violence and warfare in the Democratic Republic of the Congo, which produces nearly 40% of the world's tantalum.

Intention

To fight the horde, and sing and cry
Valhalla, I am coming

-Led Zeppelin, *The Immigrant Song*

There's a lot of strife where tantalum is concerned. It was named for Tantalus, one of Greek mythology's psychopaths; not because of his crimes, but for his punishment. Tantalum is found only in mixtures, and is very difficult to separate from niobium, so the pure metal—and the right to say one had discovered it—remained just out of reach for quite some time. Intention requires persistence.

These days, tantalum is still causing problems, but of a different sort. It is technology-critical; every type of current tech uses it. Only a very small amount is needed to make durable, high-traffic resistors and capacitors in just about all of our devices…And one of the key sources for the mineral containing it is in central Africa, where wars are being waged over who gets to own the mines. Intentions can be toxic when fueled by greed.

Though on a much smaller and less harmful scale, we can all become blinded by tunnel vision. This does us, and our aims, a disservice. Intention can certainly be destructive; but it does not have to be. Think of the world that has opened up because of tantalum-dependent microelectronics. The possibilities are mind-boggling, and it is sacrilege that people are killing in order to control it. Intention need not be corrupted. When the target is lofty, and our aim is true, we can accomplish great things.

Tantalum comes to you with a double message, and it's best to take them both to heart. First of all, go get those aspirations! Take your mark and draw back your bow, aim for the stars or higher! Make your visions live! And while you do, be kind. Be aware of those around you. Do not confuse greed with goals. Life is not a zero-sum equation, and intention is only worth pursuing when it benefits everyone present.

Tungsten

Element Number:	74
Weight:	183.84u
Normal State in Nature:	solid
Melting Point:	3695° Kelvin
Boiling Point:	5825° Kelvin

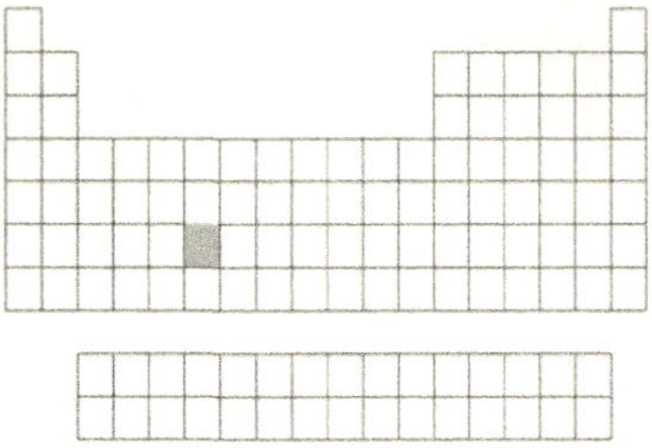

There was a mineral called tungsten (meaning "heavy stone") which Scheele, in 1781, discovered could yield a previously unknown acid, which he called tungstic acid. He thought perhaps that reducing it might create a new metal.

Two years later, wolframite turned out to be capable of producing an identical acid, and the Elhuyar brothers successfully reduced it, naming the resulting metal wolfram. That name is still used in many places, and is where the symbol for tungsten (W) comes from.

Wolfram derives from the old Latin *lupa spuma*, "wolf froth," because back in 1546 the extraction of the ore required enormous amounts of tin that would vanish during the process, as if a wolf was eating it.

Pure tungsten doesn't exist in nature. The carbide is its main use, in cutting blades and drill bits, being so hard it will penetrate solid rock, concrete and metal armor. The tips of trekking poles are tungsten carbide, as are the studs on snowmobile tracks. Surgical tools of it are prized. It is also in the tip of your favorite ballpoint pen.

It used to have applications in lightbulb filaments and in permanent magnets; it is still used in weaponry, such as armor-piercing bullets and grenades. Cars, planes, boat ballast, shuttles and turbine blades usually have a little bit of tungsten somewhere in their makeup.

Adapt

Adopt, adapt and improve! Motto of the Round Table.

-John Cleese, *Monty Python's Flying Circus*

Tungsten was used in the late 1800s to make steel magnets harder. Then in the early 1900s, it was discovered that tungsten filaments make the best incandescent lightbulbs. So durable! So bright! But sadly, incandescents are inefficient, ineffective and, with the advent of CFLs and LEDs, obsolete. This would have been a terrible blow, except that, before the demand for its lightbulbs had even begun to flicker, tungsten had adapted to being used in machining. Tungsten carbide is very very hard, maintains an edge, and is more resistant to fracturing than a diamond. This would be a good place to stop evolving, but tungsten isn't done. It is dense, cheap and has a high melting point, so it has found its way into all sorts of industries, from rocket nozzles to jewelry. It alloys beautifully, and it can associate either with violence, as when it is cast into armor-piercing bullets; or relaxation, as when it is used in fishhook weights. To be able to adapt is a skill to be prized. Sometimes we get so attached to our ideas of who we are and where we are going…that when life throws a wrench in the works, we choke. And life will do this, on a fairly regular basis. The tasks that were crucial yesterday may be antiquated tomorrow. But the skills, the talents, the fundamental way we approach problems—those will always be needed. If not for magnets, then for drill bits.

Be ready to adapt *is tungsten's message for you. Something in your world is about to change, and you have the ability to weather that change gracefully. Whatever transition is heading your way, tungsten is here to remind you that whatever makes you* you *is staying the same. All you have to do to come through this with flying colors is to know yourself; know your strengths; and know who you alloy with best, because their help may be needed. And when in doubt, remember: You can only forge ahead…if you let go of the lightbulb.*

Rhenium

Element Number:	75
Weight:	186.207u
Normal State in Nature:	solid
Melting Point:	3455° Kelvin
Boiling Point:	5870° Kelvin

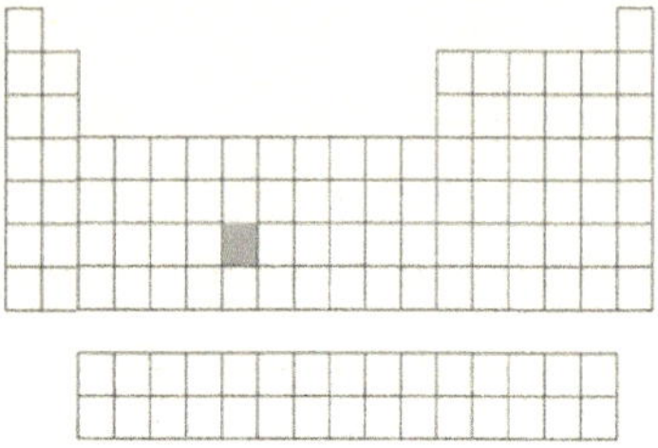

In 1908, Ogawa triumphantly declared that he had found element 43, and named it Nipponium after his country, as Nippon is one of the names for Japan.

Of course, 43 turned out to be the radioactive and synthetic technetium, and what Ogawa had actually discovered was its downstairs neighbor, element 75. His name for it was dropped because of this miscalculation, when it was discovered on purpose in 1925 by Ida and Walter Noddack and Otto Berg, who decided to name it after the Rhine river. Despite the earlier confusion, rhenium is considered the last of the stable elements to be discovered.

It took three years to extract the first gram of the metal.

Rhenium is one of the densest elements, and one of the rarest. It has very high melting and boiling points. It is extracted from molybdenite, and used in superalloys, particularly in fighter jet engine nozzles. It is unusually stable, and is used as a flashbulb filament in applications where the bulb does not contain a vacuum. It can be applied as a catalyst in creating cleaner gasoline, and its alloy with tungsten is impressive as well. Two isotopes are used to treat cancers. But, all this being said, it is very, very rare and very expensive and if you have any in your home you're probably an aerospace engineer.

75

Elusiveness

There are such a lot of things that have no place in summer and autumn and spring. Everything that's a little shy and a little rum…And then when everything's quiet and white and the nights are long and most people are asleep...then they appear.

-Too-Ticky, *Moominland Midwinter*

Rhenium is one of the rarest elements on Earth, and is the last of the stable elements to be discovered. It is very strong and hard, it has an incredibly high melting point and in fact the highest boiling point of any stable metal; if it were not so very shy and elusive and—above all—expensive, there would be hundreds of uses for it. As things stand, however, there is one industry that needs rhenium, is devoted to rhenium and claims more than half of the rhenium produced every year:

Fighter jets.

Specifically, the turbine blades in fighter jet engines require a 6% rhenium superalloy. It is as if a rare, fey creature were found in the woods, and it were only willing to engage with humankind if they gave it wings and weaponry. Sure, rhenium can do other things, and can be found other places; even in years where jet engines take 70% of the world's supply, that means 30% goes…elsewhere. X-ray tubes. Mass spectrometers. Flashbulbs. Gasoline. But, by and large, what this shy, fugitive element does best is fly at supersonic speeds. And occasionally fight.

If you have drawn rhenium, it is a reminder to draw forth that within you which is the most difficult to find. It could the strongest part of you, or the wildest, or the angriest; but it hides. It would be too exhausting to be that person all the time. But right now, you are going to have to fly and you may have to fight. Bring forward the version of yourself that is rarest and strangest. By doing so, you will inspire those around you to do the same. Alloys are stronger, and rhenium, once it has been coaxed out of hiding, never works alone.

Osmium

Element Number:	76
Weight:	190.23u
Normal State in Nature:	solid
Melting Point:	3300° Kelvin
Boiling Point:	5300° Kelvin

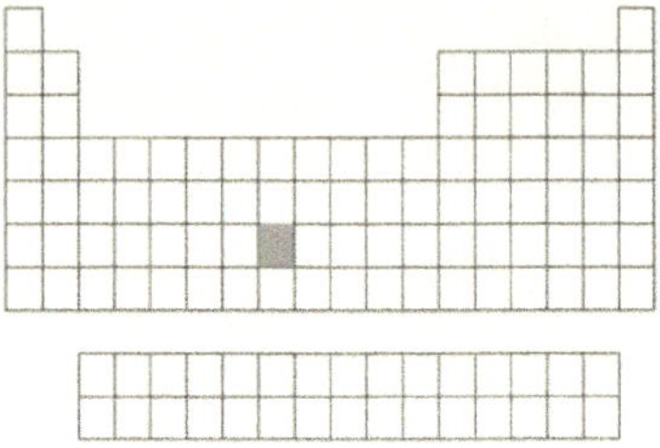

Osme is Greek for "smell," so osmium is what Tennant named a new metal he discovered in 1804. He, as well as others, were investigating the platinum ore residue left after the compound had dissolved in aqua regia, and Tennant noticed that this particular powder sublimed into a vapor that smelled of smoke and ashes. That vapor was osmium tetroxide, an incredibly toxic gas, so finding out how it smelled definitely hurt him.

Osmium was thus discovered by accident, at the same time iridium was. Two hundred years later, it's still a byproduct, used to make other metals harder.

Osmium in its pure state is a bluish metal that will always react with oxygen over time and then poison the air, so it is rarely utilized alone. It does good work in alloys, especially with other members of the platinum group, where it is protected from oxidation; and "osmiridium" has long been a favorite hard-wearing coating for phonograph needles, fountain pen nibs and electrical contacts. It's used medically to treat arthritis, and is being explored for treating cancers, having fewer side effects than some current drugs.

Every other use for the metal is hampered by its rapid oxidation rate. Its powder is pyrophoric. The tetroxide itself is used in fingerprint detection.

76

Endurance

Storms make trees take deeper roots.

-Dolly Parton

Osmium is the densest of the naturally occurring elements. It is hard, tough, and brittle. It is also very, very rare, like its neighbors, and has a unique blueish tint. Its uses are mostly quaint and archaic; a tiny dot of it on the tip of a fountain pen or a phonograph needle gives that item extreme durability. In a more modern vein, osmium is used to coat electrical contacts, again, protecting them from wear. So the heck what, we may think at this point. Yet another tough-ass precious metal in a block of tough-ass precious metals. What makes it special?

One of its more unexpected traits is, when ground to powder and left exposed to air, it will gradually oxidize; the resulting osmium tetroxide will sublime into vapor; and this vapor is toxic. Like, really, really, *really* toxic. Probably best not to grind it up in the first place.

Because the metal itself is harmless. It makes things that endure, and, though some of its applications are niche, one can still find fountain pens and record players. The demand for them is smaller, but it, too, endures. Osmium, so dense, hard and scarce, is nevertheless still here. Our own ability to withstand the ravages of time or storm can also be hard to find, perhaps. But once gathered, it doesn't just make us more stubborn or stoic; it gives us a way to preserve our thoughts. It gives us clarity of connection.

It gives us music.

Osmium is here to serve as a reminder that you have it within you to get through everything life throws at you. You really do. Look! You are still here! You are a marvel.

But it also serves as a warning: Do not allow yourself to be ground down. You can harm yourself and others when your solid state is compromised. The way to avoid this is to keep yourself warm (osmium's melting point is quite low, you don't need intense heat) and keep your boundaries firm. Draw them, if needed. With a good, sharp pen.

Iridium

Element Number:	77
Weight:	192.217u
Normal State in Nature:	solid
Melting Point:	2720° Kelvin
Boiling Point:	4700° Kelvin

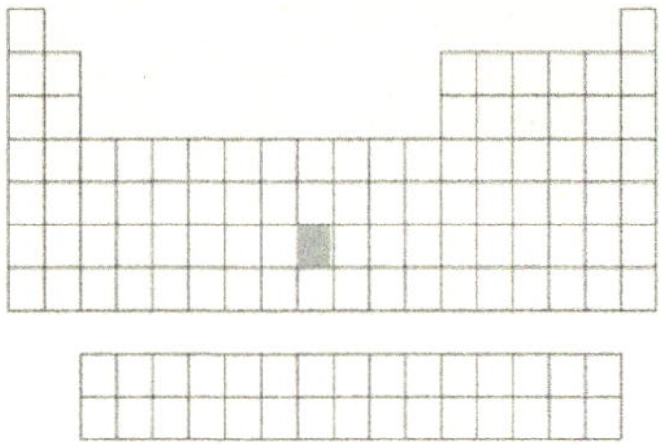

Many chemists were looking to find new metals in platinum ores, and the fact that he had more raw material to work with gave Tennant the advantage over the others. Vauquelin, for example, obtained enough insoluble residue to give it a name -- *ptene* -- meaning "winged," from the Greek; but not enough to go any further. Whereas Tennant first discovered osmium, then obtained brightly colored salts of an element which he named after the winged rainbow goddess Iris, and iridium was officially discovered. He immortalized his finding in a letter he sent to the Royal Society in 1804.

The pure metal was obtained by Hare in 1842.

Being very hard and impervious to corrosion, iridium is alloyed on spark plugs and in crucibles where strong acids will be used. It is very rare on Earth, probably because our planet's initial reserves of it sank to the core while the surface was still mostly lava.

Nowadays, almost all of the iridium available to us comes from space, where it is much more common. Meteorites can contain 4 or 5 ppm of iridium, while crustal rock contains 0.001 ppm. A thin layer of iridium within the Earth's crust is believed to be the debris from the asteroid that killed the dinosaurs, according to the Alvarez hypothesis.

Visitor

Whatever other doubt there was, or whatever wonder, Enoch had no doubt at all that this strange being was not of the Earth.

- Clifford D. Simak, *Way Station*

When the Earth was molten, iridium was common. But during the planet's cooling process, the iridium, being very dense and pyrophoric, sank to the core. We ought to have none, here on the surface. Yet a tiny amount is available for use in costly spark plugs and the occasional high-temperature crucible. Where did it come from? It came from space.

The asteroid that wiped out the dinosaurs 65 million years ago was chock full of iridium. When it crashed, it also aerosolized, and the iridium dust settled in a thin layer of clay all over the world. Smaller meteorites also contain it. If we ever need a lot of it, we will look to asteroid fields to harvest it. In the meantime, your spark plugs are probably tipped with a metal that arrived here from space and may have incidentally killed a lot of really big lizards.

There are many elements that were always here on Earth, ready to be discovered. And there are also those that can only be witnessed when we make them from scratch. But iridium alone represents a third category: That which came here from elsewhere. There's a whole massive universe out there, and it might send an ambassador to us from time to time. Let us be gracious about it. And hope they are small.

Iridium is a calling card. Expect to receive a visitor. Perhaps in the literal sense, so make sure your guest sheets are clean; but it could also be an idea or a dream that arrives in your head and takes your life's work in an entirely new direction. Be ready to receive it, and embody as best you can the courteous host.

Fair warning: The arrival could possibly be catastrophic. If it is, trust that your inner cretaceous period needed to end anyway, to make room for something else. Your planet will survive and thrive. Your spark plugs will be grateful. Your visitor is, ultimately, your friend.

Platinum

Element Number:	78
Weight:	195.084u
Normal State in Nature:	solid
Melting Point:	2042.1° Kelvin
Boiling Point:	4100° Kelvin

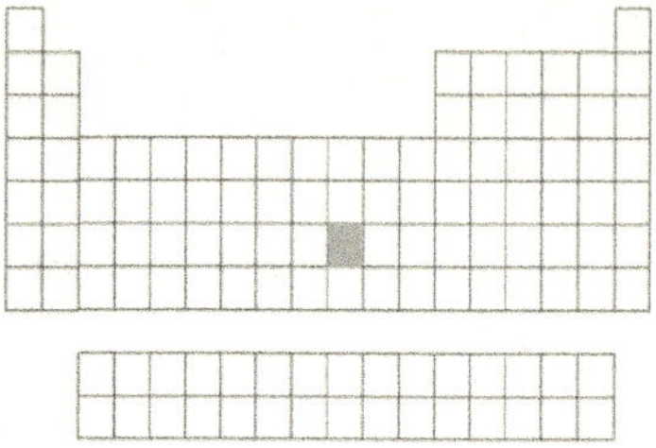

First Nations in what is now Ecuador knew and used platinum in their gold alloys for generations, but the extent of their knowledge was sadly wiped out during the invasions from Europe. Europeans first encountered the metal when explorers brought it home in the 1500s, but it was largely dismissed as an "impurity" and thrown away. It was even illegal for a while to alloy it with gold.

Platinum was examined for a while by de Ulloa, who wrote a paper about it in 1748 but then moved on to other things. Brownrigg wrote his own paper in 1750, and that excited enough interest for various chemists all over Europe to get involved. Scheffer's detailed paper in 1752 called it "white gold," and talked about how impervious it was to corrosion, and how it was almost as easy to work as gold.

Nearly half of all platinum produced today goes into catalytic converters with the other members of the platinum group. The rest is used in a great many places, from jewelry to lab equipment; from reduction of oxygen in fuel cells to chemotherapy drugs.

It has been used in the past as the prototype meter and kilogram (alloyed with a small amount of iridium), and is still used as one of the thermometers that defines the international temperature scale. It can be found in watches, electrodes and surgical instruments, and paired with cobalt to make strong magnets.

Platinum can do so much it is no wonder that it is used as a symbol for excellence. But there is no platinum in your credit card.

Mastery

"I know of no wizards."
"It is the rank beyond master in swordsmanship," Yeste said. "Bastia was the last man so designated. Long before your birth, he died at sea. There have been no wizards since, and you would never in this world have beaten him. But I tell you this: He would never in this world have beaten you."

-S. Morgenstern, *The Princess Bride*

It is rare, difficult to isolate and a "heavy" metal, whose salts are toxic; but when we want the best of the best in a wide range of applications, we go to platinum. Very expensive lab equipment is made of platinum; so are the toughest electrodes, the finest medical equipment, and the most prestigious jewelry. Its name means "little silver," and when we see how many ways it can be beaten, folded or drawn out into wire, nerds might find ourselves wondering if *Mithril* was a fantasy after all.

It is also a catalyst; crucial in the process which turn exhaust, almost magically, into plain old CO2 and water vapor. It is not alone in that endeavor; a tight little group in the center of the table are all needed for this. But they aren't called the palladium group or the ruthenium group…They are called the platinum group. Platinum is the most important, most visible and most abundant (though still rare) member. It has the grace to accept that it is welcome and necessary, in medicine, in laboratories, in industry and even at fancy parties. Its message is straightforward: When a skill is mastered, own it. Anything else is a waste of air.

To draw platinum is to embrace your mastery. Give yourself permission to admit that you're a badass. What are your skills? If someone were writing a book and they called you to ask for your expert input, what would that book be about? Own your abilities. Be confident. Do not limit yourself, do not try to make yourself smaller, do not harbor self-deprecating thoughts. Those are all brain pollution, and your friends can help you scrub it. But never doubt how valuable you are, even alone. Shiny, precious and true.

Gold

Element Number:	79
Weight:	196.96657u
Normal State in Nature:	solid
Melting Point:	1337.58° Kelvin
Boiling Point:	3130° Kelvin

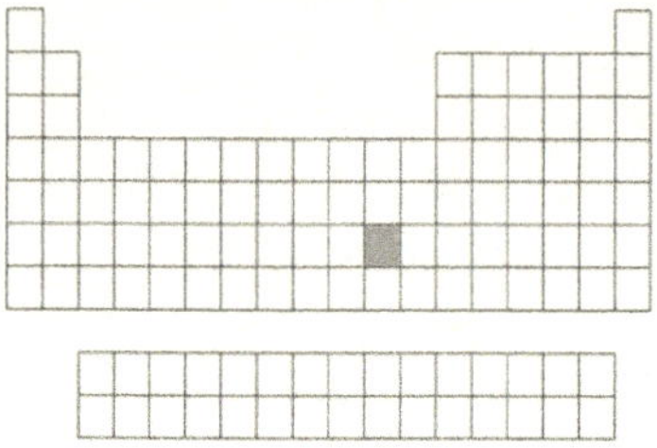

Gold artifacts have been found in Bulgaria that date back to nearly 5,000 BCE. Every ancient civilization seems to have used it, from Egypt to the Americas to China. It can be found native, it is easy to work, it does not corrode and it is beautiful. There is no mystery attached to why humans have valued it highly since before recorded history. There was even some found in Spanish caves, used by paleolithic people 40,000 years ago—although it would be a bit of a stretch to call that an artifact. It was just there.

The word *gold* is derived from Proto-Indo-European through Germanic languages and its source word meant "to gleam." The symbol is from the Latin word for gold, *aurum.*

It has been used as the basis of countless monetary systems. It has been utilized in medicine and of course in jewelry. It has endless electronic and industrial applications that have less to do with beauty or prestige, and more to do with chemical imperviousness and thermal properties.

It imparts a bright cranberry color when added to glass, and gold-plating solutions can be that same deep red or teal blue, depending on whether the solution in which the gold is suspended is an acid or a base. It has modern applications in circuit boards, and timeless importance in most religions. It causes wars. It is used as a metaphor for greatness in a hundred languages.

79

Nobility

Great heroes need great sorrows and burdens, or half their greatness goes unnoticed.
-Shmendrick, *The Last Unicorn*

Gold is an element with a lot of weight to carry. It is widely considered the loveliest metal. It is impervious to corrosion except in very rare circumstances. It can be found pure in native form, it is soft enough to be easy to work, and every civilization, ever, has highly valued it. It has always been the noblest metal.

The most ancient gold artifacts we know of are over six thousand years old. It was used to make jewelry, statuary, medicine; It has been the basis for a myriad of currencies. Our modern world has yet another reason to covet gold, not for its beauty but for its electrical conductivity. Gorgeous, noble circuitry hides in our devices. Unfortunately, it has its burdens. Gold has a bloody history. And since there are no more native gold nuggets, it must be extracted from crushed stone using a sodium cyanide dissolution process, after which the ores, filled with heavy metals and chemicals, are dumped in heaps only less environmentally damaging than nuclear waste. None of this is the fault of the metal, of course. It's not even toxic—or rather, nothing in the human body can touch it, so it is effectively non-toxic. Greed is base, but gold remains noble.

Nobility is the ability to rise above circumstances, both good and bad. It is the ability to stay incorruptible, even in the face of great provocation. It is the part of us that endures when all else is confused or polluted. Gold is reflective, ductile, forgiving. At our best, so are we.

It's a big deal to be drawing gold. The card expects you to accept the challenge of rising above whatever you're going through—or being put through. Deep in your soul, you too are incorruptible, imperishable and shiny. Bring that to the surface. Your actions and speech carry weight that you may not be aware of, so step gently. Act honorably. Speak truthfully. Embody nobility.

Mercury

Element Number:	80
Weight:	200.59u
Normal State in Nature:	liquid
Melting Point:	234.31° Kelvin
Boiling Point:	629.88° Kelvin

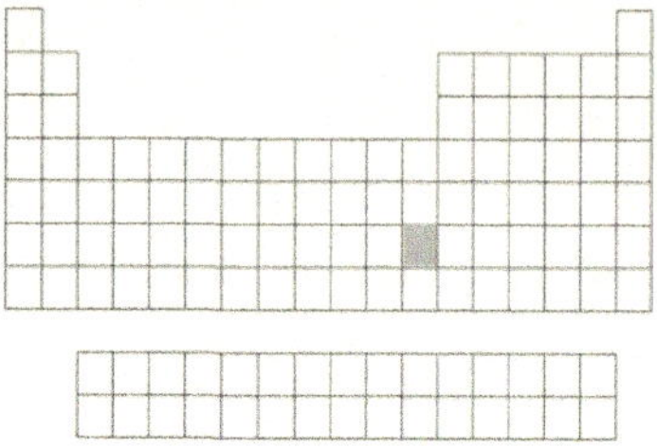

Its symbol comes from its name in Greek, *hydragyrum*, or "silver-water." And as it is a shiny liquid metal, mercury has been beloved and coveted by almost every civilization for thousands of years. It was used in ancient Chinese medicine, in paleolithic pigments, in the tombs of royalty from Egypt to Mexico. Alchemists believed it to be the key to transmutation of elements, and possibly even the key to immortality. The first Chinese emperor drank a mixture of mercury and jade to become immortal. He died. The source of the word *amalgam* is from old Latin meaning mercury.

More modern applications from the 19th and 20th centuries included hat making, dental fillings, and all kinds of -ometers used to measure everything from temperature to blood pressure.

All of these uses are being phased out, as mercury has been found to be incredibly toxic. Hat makers were driven insane by it with such regularity that "mad as a hatter" became an idiom. It accumulates in fish and shellfish, and does not go away when you cook them.

Anyone over fifteen years old who's ever had a filling has mercury in their mouth. The old bluish street lights were mercury-vapor ones. Also, it corrodes aluminum badly, so it's not allowed on airplanes. Inhaling the vapor interferes with basically every bodily system, and that statement includes industrial pollution, as mercury is one of the off-gases of coal-burning plants.

Obsession

Now I've got that feeling once again
I can't explain
you would not understand
this is not how I am

-Pink Floyd, *Comfortably Numb*

Mercury is a very strange metal indeed. It is also a very strange liquid. It solidifies at roughly -38 degrees (both Celsius *and* Fahrenheit). Throughout history it has been envisioned as the gateway to untold mysteries. Mercury is dense enough to float a bowling ball, electrically conductive and just…magical. Gaze upon it. Lose yourself in it. It flows together like thoughts or like galaxies, shining and heavy and weird…And toxic. Oh so very toxic. Humanity has found hundreds of ways to use it, and every single one has resulted in environmental contamination, or madness, or death. Or all three. Batteries, thermometers, dental fillings... Mercury can do so many things, but all of them at a terrible price. Medieval alchemists would be appalled. To them, it was the "mother metal," the key to transmutation, the seed of immortality… Mercury was everything.

It is so easy to become obsessed. We slip into it slowly, until we can no longer think of anything else. For some it is their work; for others it is addiction to some substance or other, there are so many, any one of which can eat the mind and drive us mad in the end. We must beware.

Do not look into her eyes. Do not take her hand. Mercury is a dire warning, there is something in your life that could very well become an obsession, and you will suffer if you let that happen. It may be hard to see, because obsessions have their own defense systems. Looking at yourself with clarity can be difficult, but try. Or ask a friend. Most things that will hurt you are not as literally deadly as mercury; however, getting free of an addiction is a lot harder than avoiding it when it hasn't quite taken hold. Shake it off. Walk away. Be wary. Obsession is never your friend.

Thallium

Element Number:	81
Weight:	204.3833u
Normal State in Nature:	solid
Melting Point:	577° Kelvin
Boiling Point:	1746° Kelvin

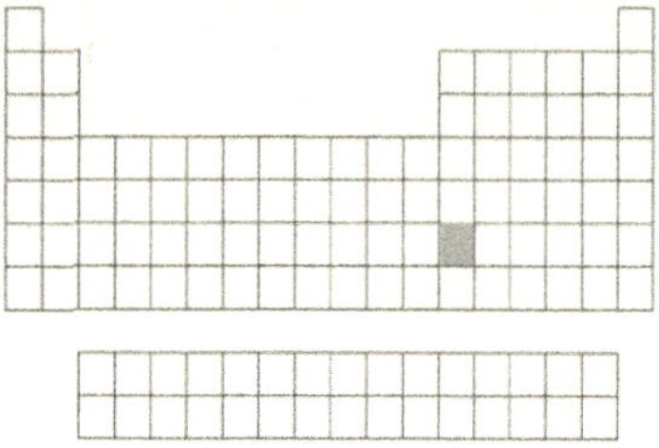

Named for the Greek "green twig," thallium was discovered independently by Crookes and Lamy in 1861, using flame spectroscopy. Crookes was working with natural ores, while Lamy was using the byproduct of his friend's sulfuric acid plant. This gave him the advantage, as he had a much larger quantity to work with, and he was even the first to isolate a small ingot of the metal; but Crookes was determined to get credit for the discovery, threw a fuss when Lamy was awarded a medal for his work, and eventually ended the debate about who was going to get the acclaim for finding the element by becoming a Fellow of the Royal Society in 1863.

He alone is credited with the discovery.

It had some uses in the 20th century, which have mostly been discontinued.

Thallium sulfate is odorless, tasteless and extremely toxic, so it was widely used as a rat poison until it was banned for this use in the 1970s after many accidental deaths. And several that were less accidental. Thallium did become a favorite murder weapon when arsenic fell out of favor. It kills within a week, and all of its symptoms can be attributed to random health problems.

Other thallium compounds are still used in optics because of its high index of refraction, and photoresistors because its electrical conductivity changes with exposure to infrared light. It can be alloyed with other metals, sometimes improving the grain. Still, it is very toxic, and should be used with caution.

Malice

"So here the new child has someone to hate with all its unreasoning mind."

Ray Bradbury, *The Small Assassin*

About a hundred years ago, thallium surpassed arsenic as the murder weapon of choice for those impatient to inherit dear old granddad's fortune. It kills quickly, and every symptom of thallium poisoning, such as hair loss, delirium and vomiting, is easy to mistake for other causes. It used to be sold as a rat poison, but it was taken off the market because, since it is water-soluble and tasteless, there were too many accidental deaths, above and beyond the various murdered granddads.

Thallium has many benign uses, in glassmaking and optics and electronics. It is showing promise as a high-temperature superconductor and is even employed as a tracer in pulmonary imaging. But in these applications, it is tightly controlled, because it is entirely too possible for it to slaughter factory workers or health care professionals offhand.

Malice can be directed at someone or something specific, which makes it easy to recognize and seek therapy. But it does far more damage when it is indiscriminate. Many of us may curse a slow barista or an idiot driver or a difficult customer…or ourselves, when we do something boneheaded. And whether we just mutter aspersions, or go out of our way to make their lives harder because they damn well deserve it…we are malicious beings in that moment. And that is very bad.

Thallium is here to caution you: Your malice is causing harm. Who is it that you hate? Why do you hate them? Are they really worth it? Or is it not a person, but a set of circumstances or obstacles that has won your unyielding dislike? Give yourself the space to admit that you are feeling this way; then remind yourself that this is not healthy or helpful, and try to direct that energy elsewhere. Thallium has other uses, remember. You can turn your energy away from malice. Make it into a window to see the world more clearly. Use it to look into your own heart. But you have to stop allowing it to poison you first.

Lead

Element Number:	82
Weight:	207.2u
Normal State in Nature:	solid
Melting Point:	600.65° Kelvin
Boiling Point:	2023° Kelvin

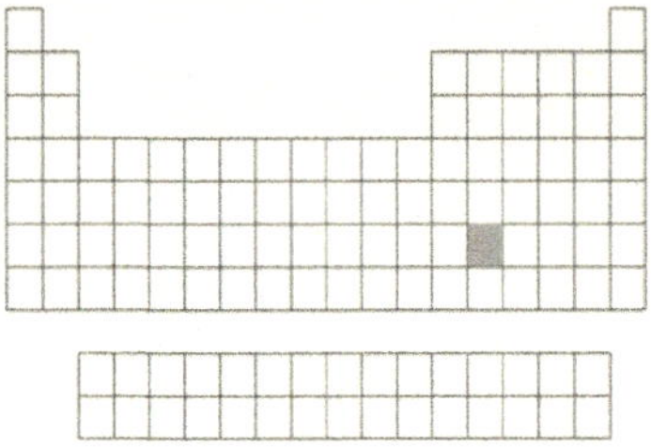

Since it exists in ores that also contain silver, it was inevitable that lead would become a commodity as soon as humans learned how to smelt. Early civilizations used it for everything from sling bullets to coins to cosmetics, but because of its dull appearance it was not really used in jewelry. It is soft enough to work cold, and its salts make nice pottery glazes. In ancient China it was used as a contraceptive, and the Romans utilized it for their pipes. In fact, their word for the metal was *plumbum*, which is where the word "plumbing" comes from.

In the 20th century, gasoline, house paint, and many other products had lead added to improve performance. Many of these proved to be unfortunate, as lead is very toxic indeed.

A low melting point paired with a high density gives lead plenty of uses still, even if many of its older applications have been abandoned. Bullets are still cast from a lead alloy. It can be found in boat ballast, diving belts, gym weights and x-ray protective screens.

Having no natural resonance frequencies, lead is used in soundproofing screens as well. Lead-acid batteries, leaded glass, and leaded bronze are all still active applications. One can still find lead solder, but not for much longer.

Harbinger

Something moved across the room, like monkeys and like shadows. The torches on the walls went out, the two clocks stopped, and the room grew colder. There was a smell of old, unopened rooms and the sound of rabbits screaming.

James Thurber, *The Thirteen Clocks*

There has always been an abundance of lead; and since it is easy to extract from ores, melt down, cast, alloy, and recycle, there have always been abundant uses for it as well. Ancient Rome invented plumbing on the strength of this metal, their name for which was *plumbum*. And we are not so different now. What can we make out of a metal that is dense, plentiful, and soft enough to be cold-welded? The question is, what *can't* we make.

The only problem with this is that lead is famously toxic. As a heavy metal, it builds up in a body or an ecosystem, and cannot be flushed or processed. It destroys enzymic functions. It attacks the nervous system. Every time it turns up, it is a harbinger first of some brilliant innovation, and second of sickness and death. Leaded gasoline, leaded paints…chuckling quietly to itself, lead has been enfolded into all sorts of industries, where it calmly ushers in ruination until it is banished.

Oh, but we aren't done with lead. Lead-acid batteries are still very popular, as is leaded glass and bullets and even radiation shielding. Like a strange old sybil, it wanders into any place it hasn't been barred from, and does harm—but more than that, it shows us how willing we are to harm ourselves. And each other. Lead's popularity is a portent of bigger things, and we would be wise to keep our eyes open.

Lead is a sign to watch for the signs. Omens and portents, strangers' conversations overheard, weird thrift store kismet. Perhaps you'll see your spirit animal while taking out the trash, or dream of a warning from your dead grandfather. Whatever. Keep your eyes and ears open. And your heart. Listen to the messages the universe is throwing at you. And if the messages conflict, follow your gut. It knows when something is pretending not to be toxic.

Bismuth

Element Number:	83
Weight:	208.9804u
Normal State in Nature:	solid
Melting Point:	544.59° Kelvin
Boiling Point:	1837° Kelvin

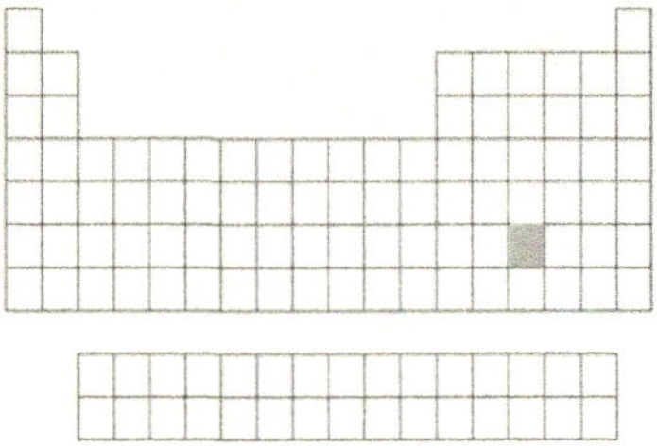

Being one of the first ten metals discovered by humans, bismuth's origin, and the origin of its name, are shrouded in the mists of time. Writings as early as the 16th century talk about it as a metal in the family of metals that includes lead and tin. Its compounds have always been used to treat stomach disorders, and its metallic form has been used in alloys. The Incas used it in bronze to make knives. Until there were antibiotics, it was a treatment for syphilis. Its name might be from the German word *wismuth*, meaning white blob. Miners in the Middle Ages called it *tectum argenti*, meaning "silver being formed."

It has interesting uses for a metal, including, still, treatments for a variety of gastric problems.

Whether you need an antidiarrheal, an ulcer medication, an antidote for cadmium poisoning or merely internal deodorant (which is indeed a thing), bismuth compounds are where you turn. It is heavy enough to replace lead in most applications where lead's poison would be a problem, and it alloys well *with* lead in other applications where toxicity is less of an issue.

Its iridescence makes bismuth a staple ingredient in cosmetics, and has since ancient Egypt. And it forms beautiful hopper crystals when cooled slowly after melting. It is effectively the last of the stable elements, since it *is* radioactive, but with a half-life of 2.01x10^{19} years.

83

Gatekeeper

"If I'm big enough to stand in the mud all this time and contemplate infinity," said the child, "I'm big enough to look at the other side of nothing."

-Russell Hoban, *The Mouse and His Child*

Bismuth stands at the threshold of a significant shift in the periodic table. Every element up to this point (except technetium and promethium) has stable isotopes; none of the ones that come after do. Technically, bismuth itself is radioactive, but its half-life is a billion times longer than the age of the universe, so it is reasonable to treat it as stable. It can inhabit both worlds: That Which is Immortal, and That Which Decays. Both sides of the gate are its domain.

Whatever else is true about the elements beyond this point, their presence is unhealthy for living things. Yet bismuth itself is not just benign, but medicinal: it is the main ingredient of several stomach medications. And if we require beauty as well as wholesomeness, bismuth has us covered. If melted down and allowed to cool slowly, it forms rainbow-hued hopper crystals of incredible loveliness. Thus it stands also at the gateway between the utilitarian and the ornate.

A good gatekeeper is someone who does not hinder passage. A good gatekeeper should have an affinity for the realms on both sides of the wall, and be able to embody whatever is needed for each traveler: Healing or warning, function or beauty. Bismuth is a very good gatekeeper.

Bismuth could point to some great change in your own circumstances, in which case he is here to see to it that you are prepared. Do you have the information you need? Step through the gate boldly, but keep your sword handy. Conversely, you may need to embody bismuth for someone else. Is there anyone in your life who is striking off into the unknown? A friend or lover or teenager who is changing careers or moving across the country or starting college? Bismuth insists that you help them. Give them the benefit of your experience. Get them through that gate in style.

Polonium

Element Number:	84
Weight:	209u
Normal State in Nature:	solid
Melting Point:	527° Kelvin
Boiling Point:	962° Kelvin

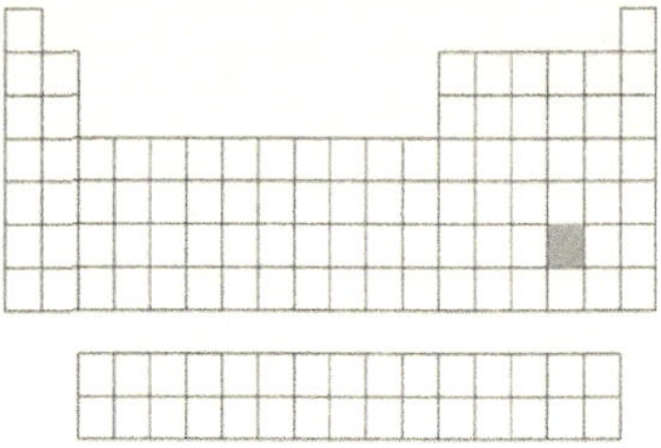

Pitchblende, also called uraninite, is a radioactive ore that Marie and Pierre Curie studied. They hypothesized that uranium and thorium were the cause of the radioactivity, but once those two elements were removed, the ore was even more radioactive than before.

This is how, in 1898, they discovered polonium, which Marie then named for her native Poland, which was under the rule of three other countries at the time and, she felt, needed publicity to fuel its fight for independence. Polonium was used as a neutron burst emitter, or "urchin," in atomic bomb triggers during the Manhattan Project, and many of its properties were classified until well after the war.

Litvinenko, a Russian agent who had defected to the UK, was poisoned with it in 2006.

Its intense alpha particle radiation means that even a small amount will spontaneously produce heat, and it has been used for that purpose in satellites and lunar rovers.

Much smaller amounts (sandwiched between gold foil to block radiation) are used in antistatic brushes in textile mills, automotive factories, photography, sheet plastic production, and other such places where static electricity buildup is a possible hazard. Polonium's alpha particles neutralize the unwanted electron charge. Because it needs to be replaced annually, and is, after all, highly radioactive, this use is being phased out in favor of less dangerous beta particle sources.

Diplomacy

Row, son. Emotion is a boat.

-Bad Lip Reading, *NFL 2020*

Polonium was notoriously once used in an assassination, that is not its usual purpose. No, despite its six month half-life, polonium has a calling in this world: Antistatic brushes.

Sandwiched between gold and silver foil, a thin strip of polonium just behind the bristles of a brush will dissipate an electrostatic charge. This may seem inconsequential if one pictures it being done to old phonograph records, but its importance skyrockets when one realizes that textile mills, automotive plants and computer hard drives all have a tendency towards static electricity buildup, which could have disastrous or deadly results. Into the breach steps polonium, smoothing the situation and wicking away the troublesome electrons. Diplomacy is, after all, the art of turning a highly charged situation into a peaceful one.

Polonium is so very radioactive, even just a gram of it will spontaneously heat up and start producing power. Which is why it traveled to the moon in the 70s, powering unmanned lunar rovers and keeping them from freezing during the long nights. Diplomacy becomes more important the farther from home we go. It can keep us energized and warm. It can keep circumstances smooth and well-regulated. It does not last forever, but it lasts for long enough.

If you have drawn polonium, bring out your inner UN delegate. Approach situations with an eye to a peaceful resolution; dispel static with concise communication. Make the people around you feel that their grievances are heard and addressed. Listen more than you talk. If you feel like you'd rather roast a bitch--we are all human, after all—remember that polonium can only do its antistatic job when surrounded by silver and gold. Gather your own nobility and clarity around you (or the friends you think embody these traits best) and then get to problem solving. You got this. You can power a moon rover, after all; these people are a cinch.

Astatine

Element Number:	85
Weight:	210u
Normal State in Nature:	solid
Melting Point:	575° Kelvin
Boiling Point:	610° Kelvin

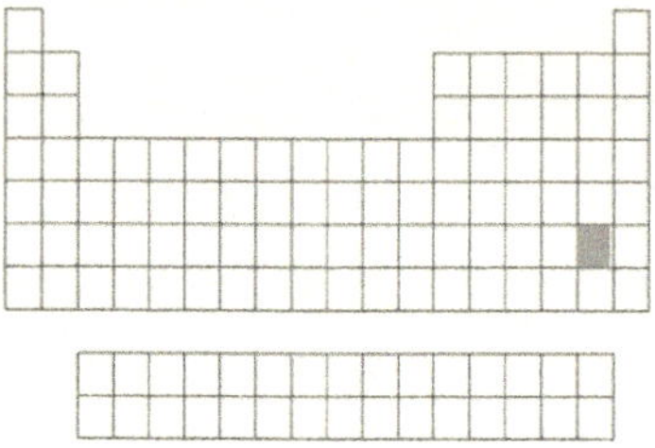

The space for astatine in the periodic table was empty for nearly eighty years. Mendeleev left a spot open for "eka-iodine" in 1869, and there it was left for decades. Because astatine is exceptionally rare and radioactive, it was not an easy discovery. There were at least seven teams between 1931 and 1940 who falsely claimed the prize. Finally, Corson, MacKenzie and Segrè went a new route, and synthesized it. They were successful, but reluctant to publish, as synthetic elements were not widely accepted as real by the chemistry community. But when it finally showed up in natural decay chains, Paneth in 1946 wrote a wake-up call, demanding that synthesized elements be acknowledged and insisting that the team who discovered it got to name it. They chose the Greek word for "unstable," *astatos*, and added the *-ine* suffix because it was a halogen.

Astatine is the rarest of the elements that occurs in nature. It exists only in certain minerals, for just under eight hours at a time, and only one atom at a time. There is estimated to be only a few grams in the Earth's entire crust at any given time. Its synthetic isotopes have possible future uses in nuclear medicine, but is otherwise not really a thing that humans interact with.

Chance

I met Thorin Oakenshield one evening on the edge of spring in Bree. A chance-meeting, as we say in Middle-earth.

-Gandalf, *Appendix A of The Lord of the Rings*

Astatine has a half-life of just over eight hours. It is occasionally part of the decay chain of uranium or thorium, but more than that cannot be said. No one knows its macroscopic properties, because more than one atom of it in one place would vaporize in its own heat. So its atoms appear and disappear, one at a time, in far-flung regions and in minerals that might be fluorescent green. Like invisible soap bubbles, here for a few hours, then pop! Then there (pop), then who knows where…

Medicine is coming up with ways to synthesize this rarest of all the naturally occurring elements, to use in fighting cancer. This is a common thread with radioactive elements. The very phrase "nuclear medicine" sends prickles up the spine, but it works. It gives a patient one more chance, and sometimes that is enough.

Chance is a funny thing. We can talk about the "laws of chance," but they are more like guidelines. Sometimes things just spontaneously occur over which we have no control. Lightning strikes, flash floods, lottery wins, finding true love…They are rare, but they happen. Chance will decide when, and to whom.

Astatine is giving you a heads-up: Chance is going to take a lead role in your life right now. Keep your eyes open for brilliant opportunities, but also for incoming disasters, it really could be either one. Astatine does not promise to deliver all good luck or all bad luck, just luck. *By the bucketload. You will not be faulted for expressing frustration against the wretched stuff that slaps you in the face, nor for celebrating the happy stuff that plonks into your lap. All your reactions are valid, but astatine's short half-life stands as a reminder that this too shall pass. So roll with the punches or seize every glorious moment; ultimately, chance is what you make of it.*

Radon

Element Number:	86
Weight:	222.01758u
Normal State in Nature:	gas
Melting Point:	202° Kelvin
Boiling Point:	211.4° Kelvin

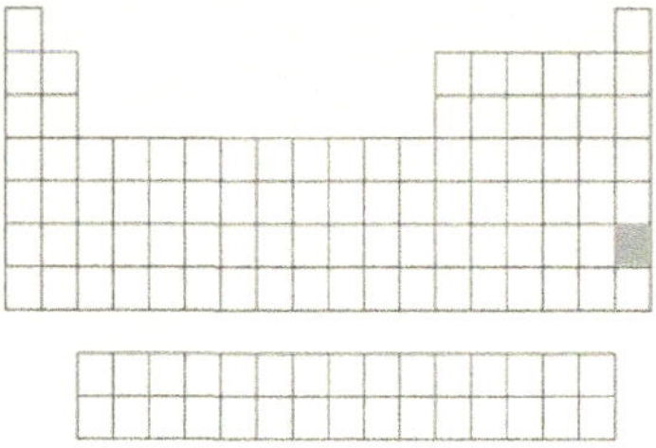

In 1899, radioactive elemental discoveries were gaining traction. The Curies had discovered four, but this fifth one was officially identified by Rutherford. He found it off-gassing from thorium, so he called it *thorium emanation*. Scientists began to observe radioactive "emanations" from two other origins, and named them for their source elements. This started to get complicated, so shorter names were suggested a few times.

It all got simplified again when all three gases turned out to be isotopes of one inert radioactive gas, which was finally named, more or less at random from the options that had been put forward, *radon:* a shortening of "radium emanation."

Long before it was recognized as an element, radon has been known simply as the mysterious cause of the "wasting sickness" miners sometimes died of. Paracelsus wrote about it in 1530. Because radon is invisible to every sense, heavy, and produced by radioactive decay within stone, the concentration of it in unventilated mines can be very high and deadly. These days, those very traits make it somewhat useful in tracking air masses, prevalence of groundwater, and other geological data. Other than that, it can be found in nuclear medicine used to treat cancer. And aside from those scant applications, humans simply (and wisely) avoid it.

86

Deception

When it comes to controlling human beings there is no better instrument than lies. Because, you see, humans live by beliefs. And beliefs can be manipulated.

-Gmork, *The Neverending Story*

Radon is a gas that happens to be radioactive. Its half-life is barely more than three days, so the fact that it is unhealthy to breathe over a long period of time shouldn't be an issue, should it? Of course not.

Except that radon is a member of the decay chain of thorium and uranium, two of the most abundant and long-lived radioactive elements on Earth. They hide in bedrock and secrete radiation. And radon gas. They always have. They will continue to for billions of years. Even when the granite has been quarried and polished and turned into buildings. It's just a Thing. It even had early 20th century scientists thinking that hot springs had health benefits *because* they were radioactive…

Radon remains deceptive, even today. It is impossible to detect with the naked senses, and there are doubtless millions breathing it in right now, all unknowing. Others have gotten their air tested and installed radon "mitigation"—that is, a hole is drilled in the basement and a fan and pipe are utilized to draw the gas upwards, to be released at roof level. It has such a short half-life, it will die soon and be no more threat. Of course, with eight times the density of air, it probably rolls back down to ground level…but it's better outside and diluted, right? Clearly we've slayed this dragon. It's gone now. Stop thinking about it.

There is nothing more poisonous than a lie. Radon in your spread is a sign that someone is lying. And these aren't those little lies like "I'll definitely start that diet tomorrow." No, these are the kind of lies that ruin lives if they're allowed to continue. Are you the victim or the culprit? Are you lying to someone for what feels like perfectly valid reasons? Stop. Or are you *being deceived? Deep down, I think you know where the lie is. Trust your distrust. Question what you're told. Defeat deception.*

Francium

Element Number:	87
Weight:	223u
Normal State in Nature:	solid
Melting Point:	300° Kelvin
Boiling Point:	950° Kelvin

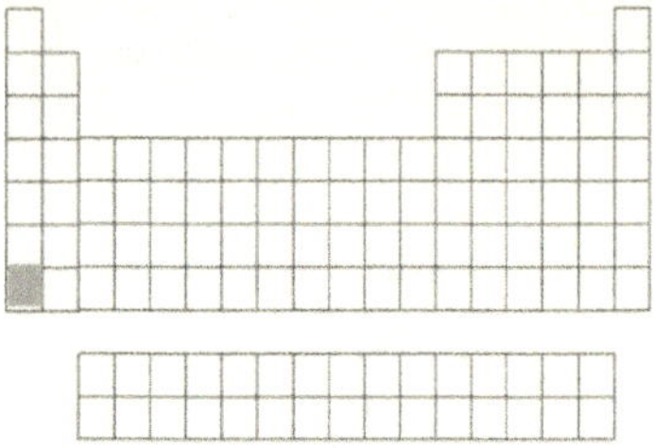

Francium is the last of the alkali metals, and it is also the last of the elements discovered in nature rather than synthesized. There were at least four false positives between 1914 and 1929, when Perey finally discovered it. She was observing the decay of a sample of actinium, noticed it was exhibiting strange behavior, and deduced that an impurity of "eka-cesium" might be the reason. Her measurements concluded that it was the most electropositive cation in the table, and she suggested *catium* for its name; but her mentor vetoed it, saying it sounded too much like "cat," and instead they settled on francium, after (what else?) France.

The largest amount ever produced was about 300k atoms. With a half-life of 22 minutes, francium has no uses beyond research.

Ephemera

I woke up wondering if we should live today as if it were the last day of our lives…because so far…it *is*.

-Abby, *Bloom County*

Of the naturally occurring radioactive elements, francium has the shortest half-life. After just twenty-two minutes in this glorious world, a francium atom will morph into either radium, radon or astatine. Its atoms are never found grouped together, either; its intense radioactivity would cause it to vaporize if it tried. No, a francium atom is born alone, and dies alone. It has no uses, not even in medicine. It exists in order to exist, and leaves without a whisper.

That which cannot be defined through observation, because it is so rare and fugitive, can at least be described by its surroundings. Francium is the final alkali metal, a family whose members all have certain things in common: They are all soft metals (cesium is so soft it's a liquid), and all highly reactive to water. In this deck they are all trolls, too, which is what gives francium her form, even as she slowly disappears.

It can be easy to forget that all things have an inherent value. Especially when discussing the elements, many of which are needed for everything from industry to nutrition…That which is ephemeral and intangible and vague seems irrelevant by comparison. But there is no comparison, not really. Francium serves a purpose just as important, if much more subtle. It reminds us that *being* is enough.

Francium whispers: This moment is all. Shake off regrets of the past and worries for the future; set down, for twenty-two precious minutes, your plans, your goals and your fears. This moment. Right here. Is all you truly have. This person, right here, is who you truly are. Tomorrow will not bring you worth, because you already have it. Tomorrow may not come at all, there were no promises made. And if it does not, be sure that the best thing you could have done with this tiny ephemeral sliver of Time…is dance.

Radium

Element Number:	88
Weight:	226u
Normal State in Nature:	solid
Melting Point:	973° Kelvin
Boiling Point:	1413° Kelvin

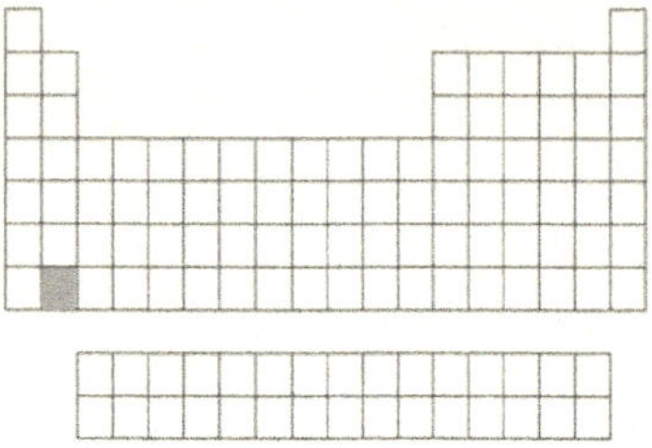

Radium was one of many radioactive elements discovered by the Curies in 1898. It was named for the French word "ray," because it emits rays of energy. By far the most notorious historical use for radium was in self-luminous watch hands and instrument dials in the 1920s, which were individually handpainted by young women who were instructed to "point" their brushes with their lips, thus ingesting radium salts. Several became sick with acute radiation poisoning, and sued the company responsible. Some gave their testimony on their deathbeds. It transpired that the company had known of the danger, and disregarded it. This became a landmark case in workplace safety law, and it also resulted in the public learning of the dangers of radiation, which had hitherto been more alluring than dreaded.

Radium exists only as a product of the decay of uranium. The modern uses for radium do not include exposing the public to it. It is very useful in several physics fields, including optical and molecular. It also can be used to treat certain cancers.

Denial

If secrecy fails, the perpetrator attacks the credibility of his victim. If he cannot silence her absolutely, he tries to make sure no one listens.

-Judith Lewis Herman, *Trauma and Recovery*

An interesting fact: When one mixes a phosphor, such as zinc sulfide, with a radioactive material, such as radium, one gets self-luminescent paint. It can be used to light up switches, instrument dials, and watches. It glows for years—until the sulfide breaks down. It will stay radioactive for much longer than that, but the level is so low, the pilots, soldiers, and other consumers are in no real danger.

Or *were*, rather. The radioactive fad peaked in the 1920s and then suddenly died. Meanwhile, the factory workers—all young women—who painted luminous watch faces were gravely exposed. They were even instructed by their employers (who knew the risk) to "point" their tiny, radium-soaked brushes with their lips. And when the radium began to kill them in gruesome ways, those employers ran a smear campaign to claim that the girls were "loose" and were actually dying of syphilis.

Denial is a formidable tool, especially backed by wealth and power. The "Radium Girls" were the most disturbing illustration of this, but there were plenty of prosperous men who were not keen to stop selling their lethal products. And there were also consumers who simply didn't want to face the fact that their radioactive belongings, so shiny and modern…were deadly. No one wants to know when they're wrong. And the more dire the consequences of continuing in denial, the harder most people will fight to preserve their illusions.

What is it that you're trying so hard not to know? You have drawn radium because you are aware, deep down, that it is time to face ugly facts. It will not be fun. It will probably hurt your pride, if not more. But there can be no healing until the door is opened and harsh reality is dealt with. It's okay to be wrong, it's okay to be misled, it's okay to have been in denial…but it's not okay to stay there.

Actinium

Element Number:	89
Weight:	227u
Normal State in Nature:	solid
Melting Point:	1500° Kelvin
Boiling Point:	3200° Kelvin

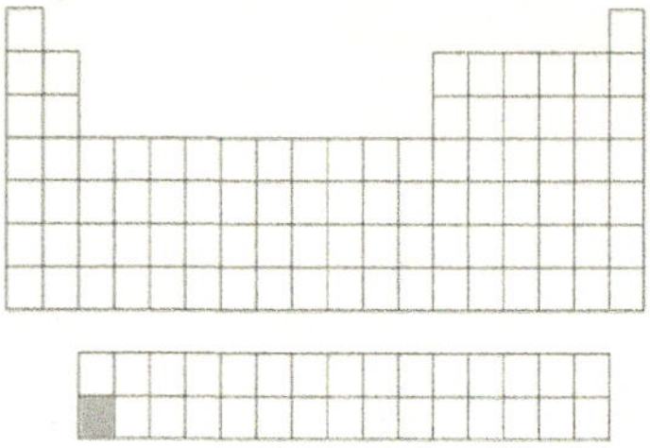

In 1899, Debierne announced the discovery of a new element, but his descriptions of it were inconsistent at best. He said it was similar to titanium, then thorium; and the chemical properties he described are actually those of protactinium, which would not be discovered until 1913.

Meanwhile, in 1902, Giesel quietly published his discovery of a radioactive element he called *emanium*, with similarities to lanthanum. And while Debierne lost interest in his substance and moved on to other things, Giesel prepared an actual sample of the metal.

Despite all this, Debierne is usually credited with the discovery, and actinium (from the Greek *actinos*, meaning ray) is the name he chose.

It is so radioactive, it glows pale blue in the dark without the need of special screens or phosphors. Actinium oxide, when paired with beryllium, is an effective neutron source, used in radiochemical investigations such as ascertaining soil density or medical imaging.

It has probable future use as a cancer treatment, and a possible use as a power source in spacecraft. However, it is far too rare, expensive and dangerous to have any widespread uses.

Stubbornness

I am both amused and annoyed that you think I should be less stubborn than you are.
-Gregory House

Actinium is the first in the actinide series, but that does not make it want to be a leader. Indeed, it apparently does not want to go anywhere or do anything. It exists in uraninite ore sometimes, but it is quite rare and unspeakably difficult for people to isolate, so mostly…people just don't. We leave it alone. We give it its space and stride briskly away.

And it could never be too briskly, either. Actinium is so radioactive, it glows pale blue without any salts or screens to assist. Its longest half-life is 22 years, and it spends those 22 years just…sitting. Maybe glowing. Sometimes taking part in experimental radiation therapy for cancer patients. But that is all. No widespread use in industry or technology for this toad. No sir. Uh uh.

Stubbornness is one of those qualities that is itself ethically neutral. It could be harnessed in support of solid ideals…or unleashed by someone choosing to be pigheaded. Stonewallers can be heroes, but they are also often petty obstructionists. Since the latter motivation exists in this world, we need the former to balance it out. Immovable objects demand irresistible forces. As long as stubbornness is a thing, it will need to be a thing.

Actinium in your spread is a sign that it might be time to dig in your heels and act stubborn; but first, examine your motives. This is important. You may think you're holding fast to your standards when really you're just being a bull-headed brat. Ask your highest self, or your spirit guides, or even just people you trust to call you on your shit; if everyone thinks you're taking it too far, graciously back off and seek compromise. But if they all agree that you do have a point, then go right ahead and stick to your guns. Glow with righteous indignation. Sit on that rock till the stars fall.

Thorium

Element Number:	90
Weight:	232.03806u
Normal State in Nature:	solid
Melting Point:	2028° Kelvin
Boiling Point:	5060° Kelvin

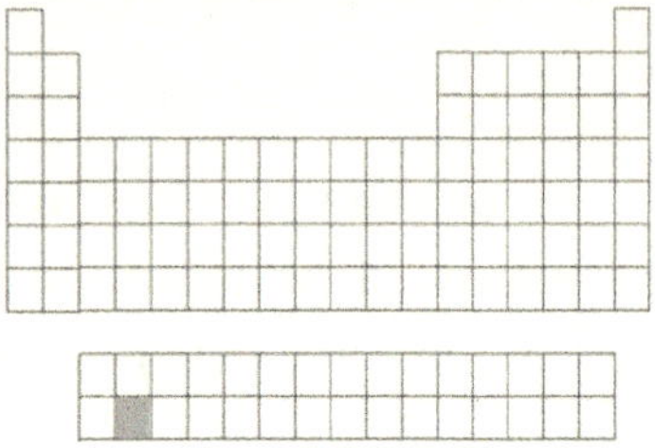

Berzelius found what he thought was a new element in 1815, but he was cautious about claiming it. He was respected and accomplished by now, but had also been embarrassingly wrong once; so he hedged his bets. But still he named it thorium, after Thor, the Norse god of thunder.

In 1824 the supposed element turned out to be yttrium orthophosphate and Berzelius graciously retracted his findings, for which piece of humility he was rewarded in 1828 when a friend sent him an unusual rock sample that *did* have a new element in it. He cheerfully reused the name he'd selected before, and thorium found its place in the table. Lely and Hamburger isolated the metal in 1914.

Thorium's historical uses started with camp lantern mantles in 1885, broadened during the radiation craze of the early 20th century to include everything from quack medical remedies to toothpaste, and then died back down again as radioactivity became better understood…although to date, one of the most stubbornly enduring uses is the lantern mantle. These are finally being phased out, and the only uses thorium has left tend to be lab-related. It makes a good alloy for high-temperature crucibles and a glass additive for certain lenses. Sure, these things are radioactive, but the risk is minimal. And thorium is being explored as a source of nuclear power as well, where the radiation would actually be useful instead of merely tolerated.

Speculation

They'll call me crazy if I fail
all the chance that I need
Is one in a million and they can call me brilliant
If I succeed

-Ani Difranco, *Swan Dive*

Sure, thorium is radioactive. A bit. Its most long-lived isotope has a 14 billion year half-life, so it's not quite as scary to be around as some of its more violently radioactive neighbors. And this is a very good thing, because there is a *lot* of thorium about. And it isn't hard to find or isolate, either. Thorium practically begs to be put into use.

When humans find an abundance of something, we tend to invent ways to utilize it. As a metal, disregarding radioactivity, thorium glows brightly when heated, making it popular as a camp lantern mantle material. It sparks well, so it can be added to welding electrodes; it can even improve the strength of other metals when employed in alloys. These are all being phased out—radiation is radiation, after all, and even a little bit is not good—but these concepts weren't terrible.

And then there's the other kind. Thorium as a quack medicinal drink turned out to be predictably deadly; its use in toothpaste was equally stupid. And in WWII much effort was made to create a thorium-based atomic bomb, a very expensive endeavor that went precisely nowhere. The jury is split on its use in nuclear power: some say it's far too difficult and wasteful, others (notably India and China) are not so sure. Prototype thorium generators are being built, even now. The speculation continues.

Speculation is the art of putting your resources where you think they will be most effective in the near future, and thorium in your spread is a sign that now is a good time to speculate. You want to be extra vigilant, however, against snake oil and "land in Florida." Speculation can backfire. So choose wisely; sidestep the projects that strike you as radioactive toothpaste; light a lantern instead, clear and bright and only a wee bit dangerous.

Protactinium

Element Number:	91
Weight:	231.03588u
Normal State in Nature:	solid
Melting Point:	1845° Kelvin
Boiling Point:	4300° Kelvin

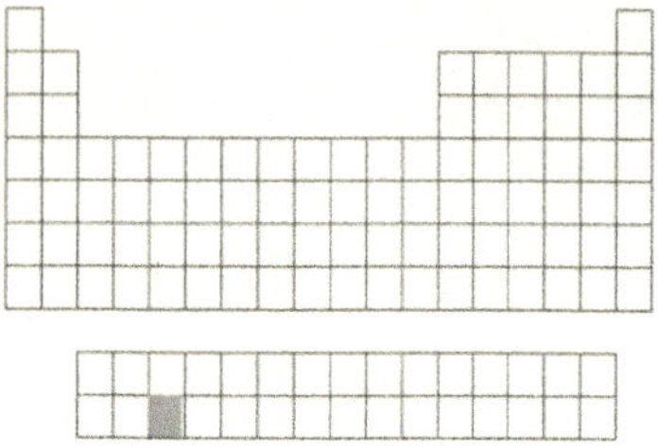

Crookes isolated protactinium in 1900, but he did not recognize it as a new element, and called it uranium-X. Later, a few different teams of scientists were studying decay chains of different isotopes of uranium, and the new element was found again: first in 1913, when Fajans and Göhring discovered a short-lived version and called it brevium; then in 1918, when Meitner and Hahn (and Soddy and Cranston) found a much longer-lived isotope. They called it protactinium, as in proto-actinium, because it decays into actinium.

A lot of credit for the discovery also goes to Ada Hitchins, a research assistant.

Protactinium is a product of the decay of uranium 235. It is considered a waste product in thorium reactors. It has no industrial applications, being too toxic to handle safely.Thus far, the only use for protactinium is in paleoceanography, because the sensitivity of modern mass spectrometers makes it possible to map the ratios of thorium to protactinium, which provides radiometric dating information.

Other than that, nothing at all uses protactinium; not nuclear medicine, not nuclear fission, nothing.

Story

Stories are like genies, Dirk thought. They can carry us into and through our sorrows…Like genies, everyone has one. Like genies, sometimes we forget that we do.
-Francesca Lia Block, *Baby Be-Bop*

Protactinium is used for…not a thing. It is so toxic and radioactive that it's not worth trying to find an application for it. Quite the reverse, in fact: It is one of the undesired products of a nuclear power plant, and is carefully removed and "disposed of" into a disused mine somewhere to sit for 33,000 years while it waits to become something less deadly.

Here's the thing, though: outside of fission chambers, protactinium is rare but, strangely, it is also *everywhere*. Tiny amounts of it—on the level of one part per billion—exist all over this planet; and scientists have found that the variations in concentration of protactinium will tell them things about primordial oceans that they could not otherwise learn. It has no practical purpose for humanity. But it tells us stories.

Stories are important. Protactinium will eventually decay into actinium. It's "proto" because it is what happens *before*. And when discussing the Great Before, all we have are the stories we tell. Our understanding of ourselves, our universe…all of it is encapsulated in stories. They are so ubiquitous we can forget that they're even there. And while they cannot feed us, they tell us what is safe to eat. They can also be so dangerous that authorities like to lock them up…but while radioactive waste isn't something we want readily available, our stories, on the other hand, belong free.

What is your story? Protactinium asks. Where did you come from? Your tattoos, your scars, the sadness in your eyes when you hear a certain song...These things matter. Your story is important. Now is the time to write a memoir, start a vlog, or at the very least buy a pretty journal and start to fill it, not with plans, but with memories. Then share them. By doing so, you empower the people around you to tell their own stories. Some of which have been locked away for a very long time.

Uranium

Element Number:	92
Weight:	238.02891u
Normal State in Nature:	solid
Melting Point:	1408° Kelvin
Boiling Point:	4407° Kelvin

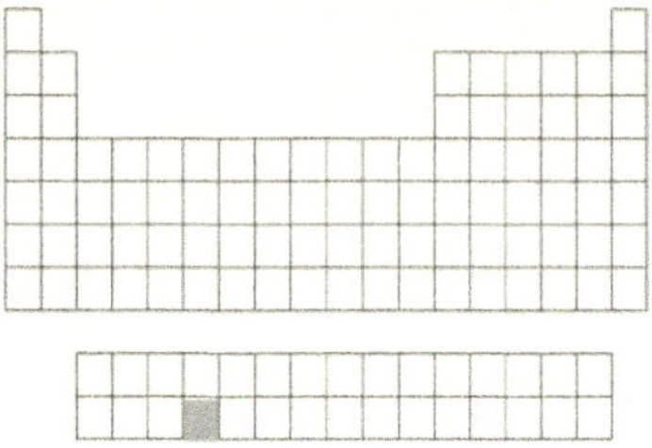

Long before it was known to be radioactive, uranium oxide was used in ceramic glazes and glass colorant. To this day, you can find yellow-orange tiles, plates and goblets that make a Geiger counter dance. But the discovery of it as an element is credited to Klaproth in 1789. He named it for the planet Uranus, which had recently been discovered.

In 1841, Péligot was the first to isolate the metal itself. In 1896, Bequerel discovered radioactivity by leaving a lump of uranium on top of a photographic plate.

Uranium was the focus of much research and development starting in the 1930s, when nuclear power and weapons became a goal for many countries. The Manhattan Project was successful on both counts. The first atomic bomb dropped on a city had a uranium core. It initially killed about 75,000 people. Many others died of radiation sickness later.

Depleted uranium is a weak alpha-emitter, so its radioactivity is less dangerous than you would expect. At the same time it is also extremely dense, giving it a number of applications. It is used in armor-piercing ammunition, shielding material, and inertial guidance systems.

When not used in weaponry, the largest single use for uranium is in nuclear power plants. The first nuclear plant to make electricity is in Arco, Idaho, the first city fully electrified by nuclear power. You can still find it in glass and ceramics, although these are getting rarer.

Uranium will never overcome its reputation as the explosive in the bomb at Hiroshima.

Remorse

He could do nothing--nothing! In all his life he had never known such bitterness of soul as the realization brought home to him.

- L. M. Montgomery, *Each in His Own Tongue*

We could have turned back during the 1930s, when the atomic bomb was but an academic theory. We could have turned back when the Manhattan Project was trying to split the atom. We could possibly have turned back after the initial Trinity test showed that this thing was a destructive force beyond any humankind had ever seen.

But once we dropped a uranium bomb on a city full of people, there was no turning back.

We have lived in the Atomic Age ever since. The threat will always be present. Fortunately, while there is a fair amount of uranium in the Earth's crust, only a fraction of it is usable for destroyer-of-worlds purposes. The rest is still radioactive, but it is merely a slight health hazard, in orange glazes and green tinted glass. Oh, and armor-piercing bullets. Because sometimes dealing death should be personal.

When we find that we have done something horrible (not just *we* as a nation, but we as individuals), it is natural to grieve—for the people we have hurt, yes, but also for the people we were before this happened. There was a sort of innocence back then. Now it is gone. Remorse is normal, even necessary; but it changes nothing.

Uranium has no upbeat message. Uranium is here because the only way to get to whatever comes next is through the valley of shame. Someone is suffering and you are responsible; but you cannot fix it. Nothing can fix it. You can only grieve. It is possible that you have drawn uranium because you started a wildfire or caused an oil spill; but it is more likely that your remorse has been triggered by something smaller. You may have irreparably hurt someone you love. You may have allowed something precious inside of you to die. Whatever. It's over now. Remorse first, then, later...healing.

Neptunium

Element Number:	93
Weight:	237.0482u
Normal State in Nature:	solid
Melting Point:	912° Kelvin
Boiling Point:	4175° Kelvin

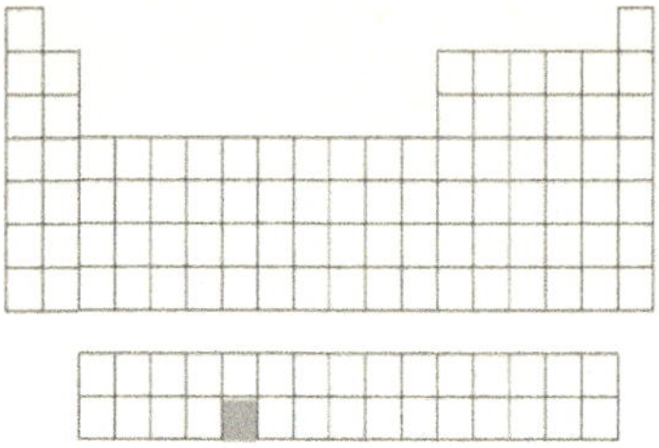

No one actually expected there to be any elements after uranium.

While in theory element 93 could exist, it was neither expected nor sought out. But in 1932 the neutron was discovered and described, and this opened up possibilities. Fermi and his team used neutron bombardment on uranium which may have produced 93, but the chemistry world was sceptical. Kimura and Nishina tried something similar and were probably even closer, but it could not be proven. Since the actinide series was unknown at the time, scientists were looking for chemical properties similar to rhenium, and this threw them off target many times. McMillan and Abelson used a cyclotron on some uranium in 1939, and by painstaking process of elimination they realized they had it. They named it neptunium, after the planet beyond Uranus.

And what does it do, after so much work? Nothing, yet. It is theoretically fissionable, but no one has made a neptunium bomb. It is a byproduct of smoke detectors, as the americium decays. The actual boiling point is unknown. The value was extrapolated from its vapor pressure.

Otherwise, its purpose is to hang around for a while and eventually decay into plutonium, for which there are several uses. One could say that its biggest use was in proving that actinides were a thing, and that there are indeed elements beyond uranium.

Patience

Rivers know this: there is no hurry. We shall get there some day.

- A. A. Milne, *Winnie-the-Pooh*

For a metal that is poisonous, radioactive and pyrophoric, neptunium is not scary. Its biggest issue is that it can form compounds with a large variety of other elements, so care is needed in its handling. Fortunately, it does not get handled much. There is too little of it. It decays after a while into plutonium-239, that sinister neighbor-element we value for its ability to go boom. We synthesize americium, which decays into neptunium, in order to get plutonium. The only purpose we assign it…is to eventually become something else. And this is fine. It knows it will transform, and does not care how long that takes, or whether we value it in the meantime.

Neptunium is the first of the transuranic elements. Everything beyond uranium does not, officially, exist in nature, but in neptunium's case (and plutonium's) this is misleading. It certainly occurs naturally, just in tiny amounts. Eventually, convention will catch up with reality and stop calling it a synthetic element; until then, neptunium will be patient. Patience is a powerful force. It may not move mountains, but it can wait until they erode. To know where we are going, but be in no rush to get there, is to take a measure of serenity. Neptunium knows that it used to be different, and someday it will be a new kind of different, but there is no need to force things along. Patience is strength.

With neptunium, you are invited to think about those things that you may be hurrying through or pushing too quickly, and take a step back. Whether you are rushing a project, or a healing, or any other of the necessary processes that make up a life...Stop. Breathe. Let it go at its own pace for a while. Let your soul grow at its own pace for a while. The things that are important enough to hasten, are the things that are important enough to take your time with. So slow down, light a candle, and chill.

Plutonium

Element Number:	94
Weight:	244u
Normal State in Nature:	solid
Melting Point:	913° Kelvin
Boiling Point:	3505° Kelvin

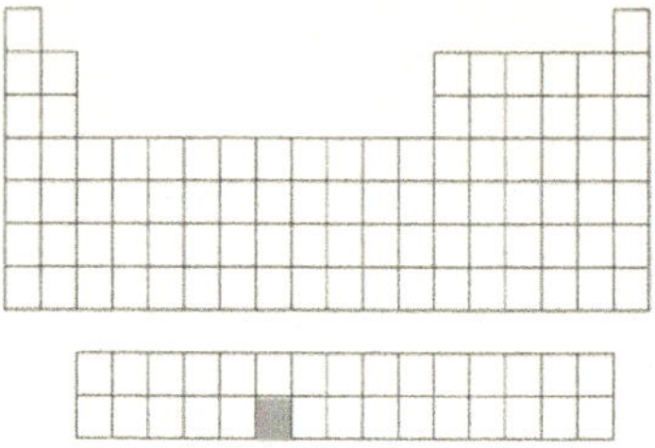

Plutonium was discovered using the same method as neptunium, but by a much larger team. McMillan suggested the name plutonium, after the planet beyond Neptune; Seaborg agreed, but gave it the symbol "Pu" as a joke, implying that it was stinky. Segrè, Kennedy and Wahl were there too.

The team wrote their paper in 1941, but its publication was delayed until after the war, because of fears of sharing nuclear secrets with the enemy. Meanwhile, a team in Cambridge discovered the same element, and gave it the same name. Being fissionable, it was a primary focus of the Manhattan Project, and the second atomic bomb dropped on a city had a plutonium core.

Plutonium's uses are in nuclear bombs, and nuclear plant reprocessing. It has minor applications as a thermoelectric power source for interplanetary rovers. It is both radioactive and a heavy metal, so it is bad for living things on all levels. Its history is a blemish on America. It was injected into test subjects in the 1940s without their knowledge, released in accidents on the edge of Denver, and covered up for half a century.

The nine countries that currently have plutonium-based nuclear weapons are the U.S., the U.K., Russia, China, France, India, Pakistan, North Korea, and Israel.

94

Control

Even while we sleep, we will find you
acting on your best behavior

-Tears for Fears, *Everybody Wants to Rule the World*

The second atomic bomb that annihilated a city had a plutonium core. The bombing of Nagasaki convinced Japan that the U.S. really was that crazy, and it ended WWII in the Pacific.

But that dubious triumph was short-lived. It didn't take long for more nations to stockpile plutonium and reverse-engineer the implosion trigger that tears nuclei apart in a cataclysmic chain reaction. Plutonium is not nearly as rare as uranium-235, the other fissionable element; but it needs absolute precision in order to cascade. Mechanical control is required if one wants to say one has The Bomb. As of now, there are nine countries who can say that. The reason they haven't yet destroyed each other (and the planet) in a fit of pique is international watchdoggery. They control one another with the threat of atomic repercussions. It isn't ideal, but it's what we've got.

When not being used to spread fear, plutonium is being sent into space. It powers exploratory rovers and satellites, which will work without maintenance for something like 80 years. So there is a bright side to all this. The exact amount of plutonium on Earth (and in space probes) is known—there never was a more rigidly controlled substance—so as long as there are just a few level heads in powerful positions, we should be okay. And when all this gets too heavy, at least we can look to the stars.

Plutonium is toxic, and so is the concept of control when taken to the extremes you are currently experiencing. Are you so desperately, rigidly in control that you are stifling the life out of yourself and the people around you? Or are you so completely controlled by someone else that you feel like your life is a prison? Either way, get the heck out of there. Make like a satellite and go far away, where you can gain some perspective. Escape this mindset.

Americium

Element Number:	95
Weight:	243u
Normal State in Nature:	synthetic solid
Melting Point:	1449° Kelvin
Boiling Point:	2880° Kelvin

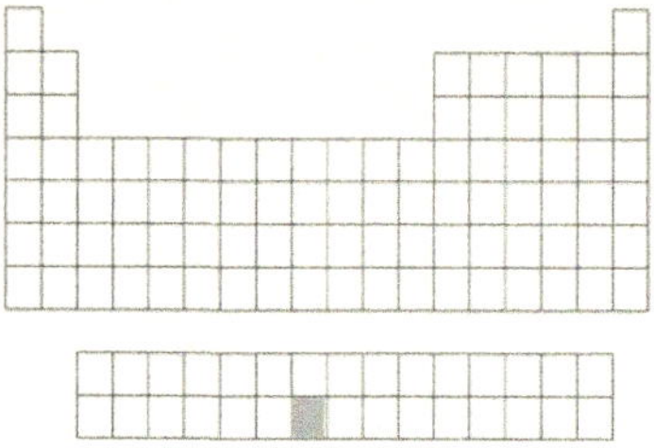

Both neptunium and plutonium were eventually shown to exist in nature, making americium the first truly synthetic element. It was probably first synthesized in nuclear tests, but was first deliberately created using a 60 inch cyclotron at UC Berkeley in 1944. Seaborg, Morgan, James and Ghiorso synthesized it, then sent it to a metallurgical lab to confirm that it was a new element.

The periodic table having been recently restructured by Seaborg himself to include the actinides, parallel to the already recognized lanthanides, the team saw that this new element was analogous to Europium, and decided it would create a pleasing symmetry to name it americium.

A microcurie of americium is all that is required to make a very sensitive smoke detector. It sends alpha particles across a small chamber in a steady, constant stream, thereby completing a circuit. Smoke particles absorb the particles and disrupt the circuit, triggering the alarm. It is much more sensitive than other smoke detectors. An American teenager tried to make a breeder reactor from old smoke alarms in his backyard, and became an Eagle Scout. The backyard was made a Superfund cleanup site.

Other uses for americium are as a neutron emitter, a starting material for other, heavier elements to be synthesized from, and as a spectrometer. Possible future uses may include a power source for space probes, and maybe fuel for them as well.

Intuition

Intuitions are not to be ignored, John. They represent data processed too fast for the conscious mind to comprehend.

-Sherlock Holmes, *Sherlock*

Americium is the first of the truly synthetic elements. From here on in, everyone will have a cyberpunk aesthetic to reflect the fact that none of them are ever found in nature. They are manmade; in americium's case, as a biproduct in nuclear reactors. It is a metal, shiny, soft and toxic, and teeming with radioactivity…but, unexpectedly, it has a calling, and should be found in every home.

Because americium can detect smoke earlier than any other system ever invented. As a tiny button sandwiched between pieces of gold foil, it emits alpha particles which zip across a simple little chamber, and are sensitive to the slightest interruption—such as smoke particulates in the air. Americium, as poisonous and radioactive as it is, has saved countless lives by sensing danger promptly, and sending out a warning. It's so quick, it's almost intuitive.

Intuition has the same job in our souls that americium has in our houses. It doesn't reason, or work things out; it cannot give you detailed instructions or tell you *why* something is bad. But when we are contemplating choices, and one of them makes the *beepbeepbeep* noise in the back of our brain, we will always be wiser to choose the other thing. Ignoring one's intuition is as reckless as ignoring one's smoke detector. Better to pay attention.

Americium wants you to listen to your intuition. It is a simple, common message, but sometimes easier said than done. So much tends to get in the way, and you may find anxiety twisting your stomach at the thought of every option. So slow down, breathe deep, quit overthinking and rationalizing and just listen for the warning beepbeepbeep which will, at the very least, point you away from bad choices. Beyond that, americium cannot help much. But it can, if you listen, at least keep you out of the fire.

Curium

Element Number:	96
Weight:	247u
Normal State in Nature:	synthetic solid
Melting Point:	1613° Kelvin
Boiling Point:	3383° Kelvin

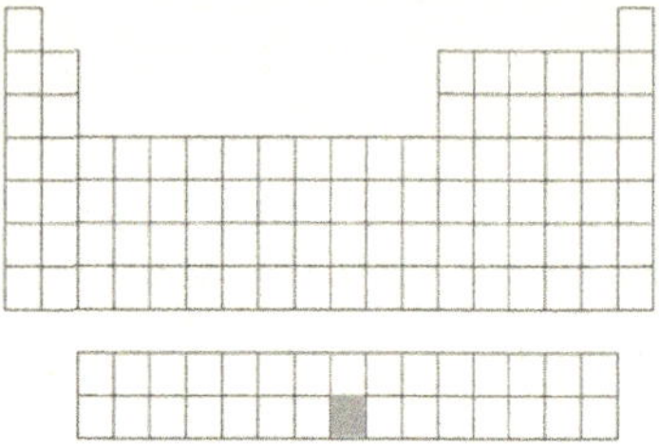

The third transuranium element to be discovered (although it is the fourth numerically), curium was synthesized by Seaborg, James and Ghiorso in 1944. They had such a hard time separating it from what would later be known as americium, that they joked that those two should be named *pandemonium* and *delirium*. Still, by using the 60 inch cyclotron at Berkeley, and sending the result to the metallurgical lab in Chicago to confirm, the Berkeley team did successfully isolate curium, named it for the Curies as its lanthanide twin was named for Gadolin; and then sat on it, because the war was on.

Curium and americium were declassified in 1945, and officially acknowledged at a November conference; although Seaborg leaked the information five days early on a radio show for children, called Quiz Kids.

Curium-based alpha particle x-ray spectrometers have gone to space with several of our probes and Mars landers. They are used to determine the composition of rocks and soil. Being one of the most radioactive isolatable elements, it does not have a lot of uses aside from that. It can be made into heavier elements, and it has potential as a power source on spacecraft.

And that is all.

Incidentally, when people talk about "the Curies" and their many discoveries, but only Marie dying of radiation poisoning, that is because her husband Pierre died at age 46 when he walked absentmindedly into traffic and was run over by a carriage.

Exploration

It'd be lovely if we could take every piece of advice we get and actually understand it right away, but that's just not how we function. We are iterative. We are iterative explorers.

-Adam Savage

Marie Curie was a powerhouse of intellectual exploration. Together with her husband Pierre, she spent years in a lab, discovering polonium (which she named for her native Poland), and radium; fearlessly researching, prying secrets of the universe out of substances that glowed.

This was before radioactivity was known to be deadly, of course. In fact, this is, in part, how it *did* become known to be deadly. While their research was invaluable, Marie died at age 66 of radiation poisoning. Her notebooks are lethal to this day. Still, two Nobel prizes, the distinction of being the first woman to hold a professorship in physics, and a powerful legacy all remain hers. Those who venture first into dangerous territory may meet a hard end, but we should still and always honor them.

Curium was first created in a cyclotron in 1944, ten years after Marie's death. It has no practical uses, except that its intense radioactivity means that it can provide power to space probes. So the element named for the queen of exploration, is also known for striking out into darkness; fearless, whatever it may find there.

Strap your boots on and grab a knapsack. Curium is a sign that now is the time to explore beyond the edges of the maps you know. Get out there and see the world, discover new things, pry secrets out of hidden caves and find knowledge no one has ever found before. If all of that sounds like too much work, then at the very least learn things you yourself did not know. Fall down an internet rabbit hole and come back with fresh insight. Go to the library and research a topic, just for the pure fun of it. Keep an open mind...but stay away from things that glow.

Berkelium

Element Number:	97
Weight:	247u
Normal State in Nature:	synthetic solid
Melting Point:	1259° Kelvin
Boiling Point:	2900° Kelvin

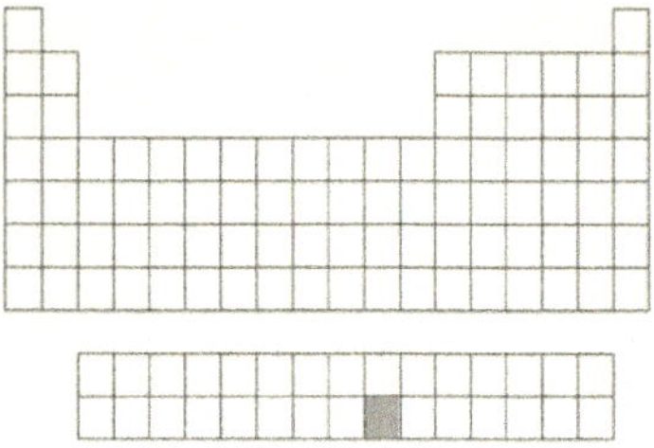

Seaborg, Ghiorso, Street and Thompson had been once again using the 60 inch cyclotron at Berkeley to attempt to create ever heavier elements. When they discovered element 97 in 1949, it needed a name. Since its lanthanide homologue, terbium, was named for a town, they proposed berkelium.

In 1958, a method was developed to create berkelium in a nuclear reactor by bombarding smaller atoms with radiation. This is how it is done today.

Berkelium's only uses are in research. It is a stepping stone for synthesizing superheavy elements, and in the other direction, for observing radioactive decay as it slowly transforms into other things. Unlike its neighbors, it is not a potential power source, cancer treatment or fuel. Very little of it has ever existed, or is ever likely to.

Intermission

I hope whatever you're doing,
you're stopping now and then
and
not doing it at all.

- Michael Pollan, *How to Change Your Mind*

Berkelium does not exist in nature. It was first synthesized in a cyclotron in 1949, and it's also a product of nuclear fission, so it can turn up in fallout and spent reactor rods. But only the tiniest amount. Some say that probably about a gram has been made over the last 60 years. That is not a lot to work with.

At least it will stick around for a while; berkelium's half-life is just under 1,400 years, so it isn't gone in a blink, like some. But it has no purpose in industry or medicine or biology or anything else, beyond pure research. Physicists can turn it into even heavier elements by bombarding it with other particles over the course of several months. They can also observe its slow decay, from berkelium to americium, then plutonium, then uranium, and on through the steps of shedding alpha particles until it drifts to a stop as lead, after something like 80 million years. So, whether it's growing or shrinking, berkelium's only identity is as a rest stop.

There's no shame in that. A pause can be a valuable thing, especially in this era of rushing around, of trying to keep pace with a world that is always in a big old hurry. Sitting down and doing nothing for a spell is almost an act of rebellion; taking an intermission can recharge our batteries while disrupting the status quo. Doing nothing *is* doing something after all.

Stop what you're doing. Put up your feet for a bit. You have drawn berkelium because it's time to pause in your busy busy life and just chill for a while. Whether you take ten minutes off, or a week, is up to you, but you must be thorough. No pretending to rest while you're actually worrying. No stressing out on the sly. Just...rest. Breathe. Savor the intermission.

Californium

Element Number:	98
Weight:	251u
Normal State in Nature:	synthetic solid
Melting Point:	1173° Kelvin
Boiling Point:	1743° Kelvin

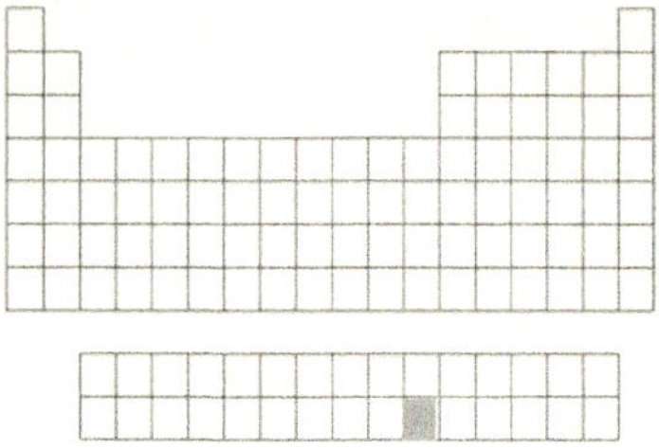

As is typical in this part of the table, the crack team at UC Berkeley consisting of Seaborg, Street, Ghiorso and Thompson synthesized californium using the 60 inch cyclotron. This particular experiment in 1950 yielded only about 5,000 atoms of the target element, which lasted for only 44 minutes.

Still, that was enough to name it. The 'twin' lanthanide had been named *dysprosium*, meaning "hard to get at," and the team was not really excited about finding another Greek phrase to translate, so they broke with the informal tradition and called it californium; their rationalization being: Well, for early pioneers, California *was* "hard to get at."

Being a very strong neutron emitter, californium's uses all reflect this. It is a startup neutron source for nuclear reactors; it can be a treatment for certain brain and cervical cancers.

Neutrons can go where other observational tools can't, so californium is used in cement and coal analysis, fuel rod scanners, neutron radiography, and other detection applications. It has even recently been used to transmit wireless data.

Despite the array of uses, californium is still very rare, and very radioactive. About a quarter of a gram is produced every year, and only two places produce it.

Perception

Images of broken light that dance before me like a million eyes
they call me on and on across the universe

-The Beatles, *Across the Universe*

So. Let's talk about neutrons.

Neutrons are the part of an atom that carries no charge. Electrons are negatively charged, and when they move around they follow certain rules; protons are positively charged and likewise have to stay in line. But when a neutron gets separated from its home nucleus, there is nothing to stop it barreling around like a tiny rogue planet and playing merry hell with anything that gets in its way. If it gets caught by another atom, it destabilizes it as a matter of course, turning it temporarily radioactive. Neutrons are chaotic little bastards.

And of all the elements, the one most known for neutron emission is californium. Every microgram of californium emits 139 million neutrons per minute (if it's fresh). This renders it extremely dangerous, so it's a good thing there is none in nature; humans make it in particle accelerators and nuclear reactors. On purpose. Because neutron emission may be terribly dangerous, but it can also allow us to see through solid matter--even things that would be opaque, or at least inaccessible, to x-rays. Firing a stream of neutrons into a steel box will tell beyond question if there is, say, a bomb in that box; firing them into a lump of ore will reveal if there is gold in that ore. It can allow us to see through anything, in fact, and all we need is highly powerful, highly portable californium.

Californium in your spread means that your powers of perception are exceptionally strong right now. Whatever you focus on, you will get a clearer idea of the truth of it; even those things that are being hidden deliberately are, to your eyes, as obvious as a kitten hiding behind a daisy. Looking into the distance, into the future, into history, into the soul...as far as californium is concerned, it's all the same. All equally transparent. And, once it has been observed, only briefly radioactive.

Einsteinium

Element Number:	99
Weight:	251u
Normal State in Nature:	synthetic solid
Melting Point:	1133° Kelvin
Boiling Point:	1269° Kelvin

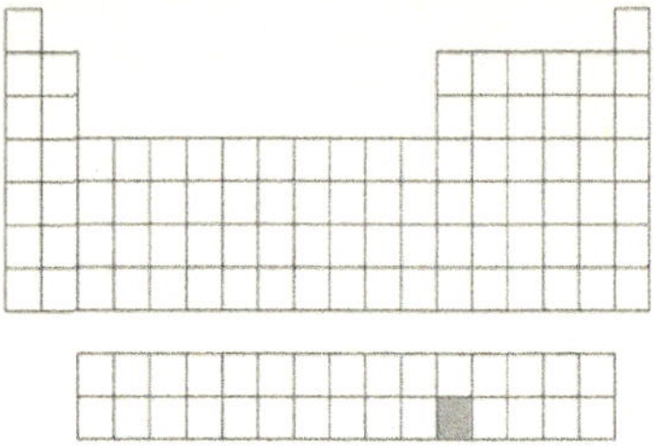

200 atoms of element 99 were recovered from the fallout of the first thermonuclear bomb test in the South Pacific in 1952. After the *Ivy Mike* test was complete, planes were flown through the airspace with filter papers attached; and, later, radioactive material was gathered from the ground as well. The analysis was carried out at Berkeley by Ghiorso and his team, but Cold War tensions meant they had to delay publication.

Meanwhile, they, and other labs, were synthesizing elements 99 and 100, by less violent means. The Berkeley team was given naming rights when it became clear that they had been first, after all. Ghiorso suggested *einsteinium* and *fermium*, to honor two science icons. Both men died within the year leading up to the public announcement.

Einsteinium has no uses, other than to sometimes be a stepping stone to superheavy elements.

Creativity

"Creativity is seeing what others see and thinking what no one else ever thought."
- Albert Einstein

Welcome to the neighborhood of the periodic table where the elements have no practical applications whatsoever. At all. Period. They are minor products of nuclear fallout or created in particle colliders, and exist purely for the sake of knowledge. They are all radioactive, synthetic, most of them are metals, and really the only interesting thing about any of them is their name.

Einsteinium is a nice place to start, if we are to focus on nomenclature. It's like training wheels; everyone knows who Einstein was, so it feels like a familiar face in a strange airport. His theory of relativity changed the world, he was a founding father of quantum mechanics, and "imagination is more important than knowledge" is one of his most famous sayings. Speaking of fame, he had it. During his life, and to this day, he was and is the most celebrated scientist in the world. But it is not so much his accomplishments that keep his memory alive. Nor is it his body of work, which got a bit bogged down in his old age with dead ends and quaint obsessions. No, what he is chiefly remembered for is his creativity; his oblique approach to the universe that allowed him to see behind the curtain of experience, and describe what he saw there. Only a creative mind could conceive the movement of light, or recognize that mass and energy are linked. Only a creative mind could solve as many puzzles as he did.

Einsteinium is in your spread because you have a problem to solve, and a conventional approach to it is not working. But the good news is, now is a particularly good time to get creative. Think sideways. Turn the issue on its head. You, too, have a creative mind, and the more you use it, the stronger it will get. So whip out a blackboard, muss up your hair, and brainstorm some weird solutions, the weirder, the better. One of them will be the one you need. As sure as $E=mc^2$.

Fermium

Element Number:	100
Weight:	257u
Normal State in Nature:	synthetic solid
Melting Point:	unknown
Boiling Point:	unknown

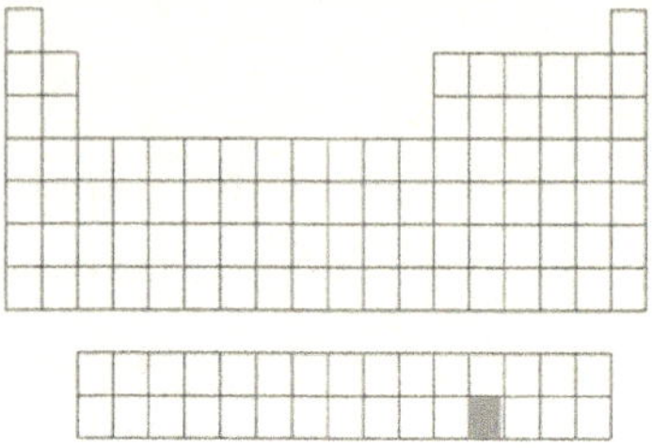

It took two months for fermium to be found in the contaminated coral beds from the *Ivy Mike* hydrogen bomb test. It only has a half-life of a day, but it is the first member of the decay chain of an isotope of einsteinium, and when they found it, they concluded that it could only be element 100.

Publication was delayed due to Cold War unpleasantness, and in the meantime the lab at Stockholm announced their discovery of element 100 through ion bombardment. However, they allowed seniority to go to the Ghiorso and the Berkeley Lab team, who named 100 after Enrico Fermi. The name was announced at the same time as einsteinium.

The most stable isotope has a half-life of 100 days. Even so, there are no uses for fermium outside of research.

Action

Running, jumping, climbing trees, putting on makeup while you're up there.

-Eddie Izzard

There have been many great minds that broke the mold in theoretical physics, and many others who have pushed the boundaries of experimental physics. But it's pretty rare to find a brain that can do both, and such a one was Enrico Fermi. His elemental namesake, like the others in this part of the table, is synthetic, radioactive and lacking in practical applications; but it can serve, at the very least, as a reason to learn something about the man himself.

His contributions to probability theory were so pivotal, subatomic particles which have a particular spin are called fermions. He was central to the development of nuclear fission and was able to predict the yield of the Trinity test. He was no monolith, either. All of his work was built on the discoveries of others, or happened in powerful partnerships. Physics is a team sport, and Fermi was an excellent player. A man of action in other ways, as well. When the Fascists made laws in his native Italy that put his Jewish wife in danger, he immediately emigrated to the U.S., where he put his brain to work for our war effort. The most famous legend about him is that he and his team created the world's first sustained nuclear chain reaction…in a squash court. Really. They had the theory, the math, the enormous quantities of graphite for neutron absorption, and the respectable quantities of uranium to fission…but they were having trouble getting the funding for a suitable location. So they decided that the University of Chicago racquetball court would do. And it did. The experiment was a success. History was made. Because Fermi took action.

Fermium in your spread is a sign that it is time for you to leap into action. You have all the tools and knowledge you need, so do it! Do the thing! Make a move! Put doubt and hesitation aside, but stay alert to avoid pitfalls. Make a plan that is bold and ambitious, and then follow that plan. Collect your team. Break assumptions. Use whatever space you can find. Take action.

Mendelevium

Element Number:	101
Weight:	258u
Normal State in Nature:	synthetic solid
Melting Point:	unknown
Boiling Point:	unknown

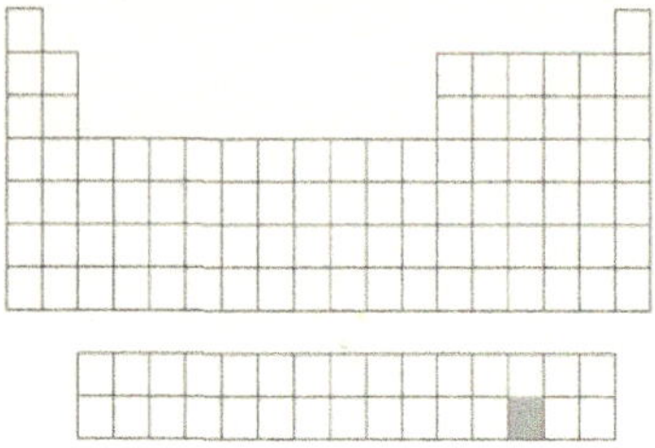

Element 101 was the first to be synthesized just one atom at a time. The Berkeley team created it, this time consisting of team lead Thompson as well as Ghiorso, Seaborg, Choppin and Harvey. In 1955 they put together some rough calculations involving what would happen if einsteinium were bombarded with alpha particles, and what the yield was likely to be. They needed an upgraded cyclotron first, so Seaborg wrote the funding proposal; meanwhile, a lump of plutonium was irradiated for a year to transmute it into einsteinium, with a 3-week half-life. They spent one of those weeks purifying it, then commenced the experiment. It was a scramble, but their results were exactly what they'd hoped.

They honored Mendeleev, the father of the periodic table, in the naming; this irritated some people, who didn't want a Russian to be so honored during the Cold War. The team ignored them.

Mendelevium has no uses outside of the study of heavy elements.

Trust

I saw in a dream a table where all elements fell into place as required. Awakening, I immediately wrote it down on a piece of paper...

-Dmitri Mendeleev

None of this would be, if not for Dmitri Mendeleev.

This table, which so elegantly lays out the qualities and relationships of every element in the universe, is his achievement. Others had postulated some sort of organization for the known elements, but since only about fifty or sixty had been discovered at the time, those theories were not really taken seriously. No one could see the whole in the parts.

Not that Mendeleev's table was seen as the mic drop that it was in 1869. As is proper in scientific circles, people argued and harrumphed and tried to disprove it, to the point that it took 37 years for someone to suggest that he be awarded a Nobel prize for his achievement—which was balked because the discovery was too old now to be celebrated.

Also, the man made enemies easily. He said scathing things about his intellectual rivals, and his messy divorce scandalized his staid Russian Orthodox contemporaries. Still, his "periodic system" has borne out; the attributes he predicted for three undiscovered elements, turned out to be dead-on; and his unswerving trust in his own vision has been vindicated. If a Nobel prize was withheld from him, he got a better tribute in an element namesake, a much rarer honor. Mendelevium is a radioactive and synthetic actinide, with no uses, sure, but with plenty to talk about. Mendeleev was a legend.

Trust is the watchword of mendelevium. I almost put "faith" on this card, but faith is a loaded word and it too often implies blind *faith. But to emulate the best things about Mendeleev, you must trust your* self. *Trust your vision. Trust your instincts and your dreams. To fall into the unknown because that is the direction your heart pulls you—that takes courage. And ignoring the clamoring voices of cynicism, outside and in, that takes focus. And when courage and focus are guided by the shining power you have within, that is the trust that can change the world.*

Nobelium

Element Number:	102
Weight:	259u
Normal State in Nature:	synthetic solid
Melting Point:	unknown
Boiling Point:	unknown

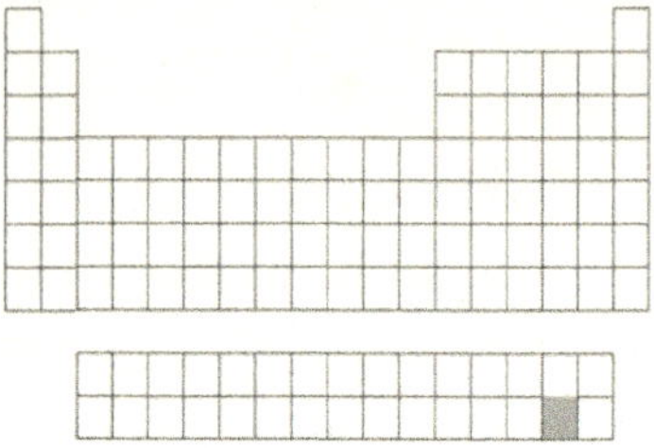

In 1957, the research team in Sweden announced that they had achieved element 102 by bombarding a curium target with carbon-13 for 25 hours in half-hour intervals. Berkeley tried to recreate their findings, and was unsuccessful. Later, it turned out that the Swedish team were mistaken. Physics teams in Berkeley and in Dubna, USSR continued to try and identify 102, with varying levels of questionable success. Berkeley was pretty sure they had it in 1959, and adopted the name that the Swedish team had proposed, nobelium. Though the name stuck, there are still doubts surrounding their conclusions. The isotope they were aiming to synthesize had a half-life of three seconds, so it was not ideal for observation. The first incontrovertible claim of its detection came from the JINR in Dubna in 1966.

Nobelium was named after Alfred Nobel, the philanthropist whose legacy is the Nobel Prize. There are absolutely no uses for nobelium, outside of heavy element research.

Generosity

The meaning of life is to find your gift. The purpose of life is to give it away.

-Pablo Picasso

Nobelium has nothing but a name, so let's talk about Alfred.

Alfred Nobel was…a lot. He was an inventor, engineer, chemist, businessman; maybe a war profiteer, but also a philanthropist. He filed 355 patents in his life, most notably for dynamite, which he invented because he wanted to render nitroglycerin safer for use in mining and demolition. His investment in his brother's oil company made him rich, his own munitions factories made him richer. None of this appeared to bother him until his brother died, and some European newspapers, confused about which Nobel had snuffed it, published scathing obituaries which referred to Alfred as The Merchant of Death. According to legend, he was appalled; and decided to restore his legacy.

So every year since his death, five cash prizes are awarded from his still vast fortune, to whomever in the world has done the best work. These are in physical science, medical science, chemistry, literature and "fraternity": The Peace Prize. One can certainly say that this repaired his legacy; his name is synonymous with international goodwill, individual achievement, and generosity.

He could have just hired publicists to rebrand his image, but instead he decided to go the eccentric route. Some say it can't count because he did it to look good…But none of us gets to decide how others share their gifts.

It is time to be charitable. Whatever your fortune is, share it. Whether you are rich in love or knowledge, compassion or motivation—or even money--whatever you have plenty of, spread it around. Give it away. Choose those who you think deserve it most, by your own qualifications. You can be loud and unexpected; goodness knows Nobel didn't make his bequest anonymously or with any collaboration. Use your own best judgment, and remember: No one gets to tell you how to be generous.

Lawrencium

Element Number:	103
Weight:	262u
Normal State in Nature:	synthetic solid
Melting Point:	unknown
Boiling Point:	unknown

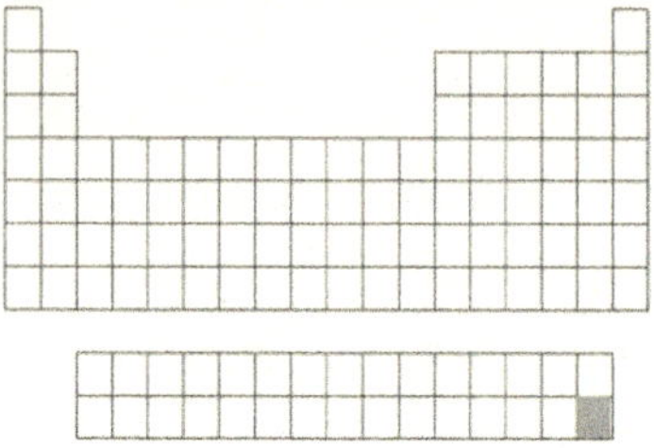

The Berkeley team first attempted to identify element 103 in 1958, but their results were inconclusive and their target material was destroyed in the process. Fast forward to 1961, when Ghiorso, Sikkeland, Larsh and Latimer tried again, this time by bombarding californium with boron. They claimed success, and published their results.

The Dubna team was skeptical, and in 1965 they began their own experiments to produce 103. Their results were arguably just as accurate, but the decay rate was different from the Berkeley numbers. Later experiments confirmed that everyone had been correct, they had merely found different isotopes. The two labs are now given joint credit for the discovery.

Lawrencium is the last of the actinides. It was named for Ernest Lawrence, the inventor of the cyclotron. It has no uses.

Drive

We are not idle. We don't know how to be. And we will not let this unique moment of stillness turn into inaction.

- Marisa Lenhardt Patton, *Death Guild Thunderdome*

This is the final actinide, which only exists officially and has no real traits to discuss. But it does have a name. In the early 20th century, science discovered that the best way to explore an atomic nucleus was to accelerate particles at each other, using a series of magnets, and record their collisions. The only problem was that one needed an enormous amount of power, as well as a very long run-up to get the particles up to speed; and land, especially in a straight line, is hard to come by.

Enter Ernest Lawrence. One fateful night he doodled on a napkin, as is traditional, a design for a device that could accelerate particles between the poles of electromagnets *in a circular motion*. When Lawrence built the world's first "cyclotron," it worked. At four inches in diameter. The forerunner of the new era was the size of a small donut. As soon as it worked, Lawrence got the funding for a larger one. When the money was tight, he got creative; he found the magnet he needed for the mark 3 in a junkyard. It would have cost thousands of dollars new. As soon as a thing worked, he started planning the next iteration. And the cyclotrons grew, and grew…Lawrence was driven in all that he did. Particle acceleration was his baby, but he applied his mind also to nuclear weapons, as well as cures for cancer. It was all the same to him: Science. Invention. Research. Discovery.

Drive.

There is something tugging on your attention, that wants your focus and your time. Lawrencium is a sign that you would do well to embrace that thing. Whether it's an art project, a room reorganizing, a life plan or a math problem, this card is your carte blanche to pick it up and run with it. Make it happen. Chase your goal, achieve your goal, and celebrate with a donut while you dream up the next thing.

Rutherfordium

Element Number:	104
Weight:	261u
Normal State in Nature:	synthetic solid
Melting Point:	unknown
Boiling Point:	unknown

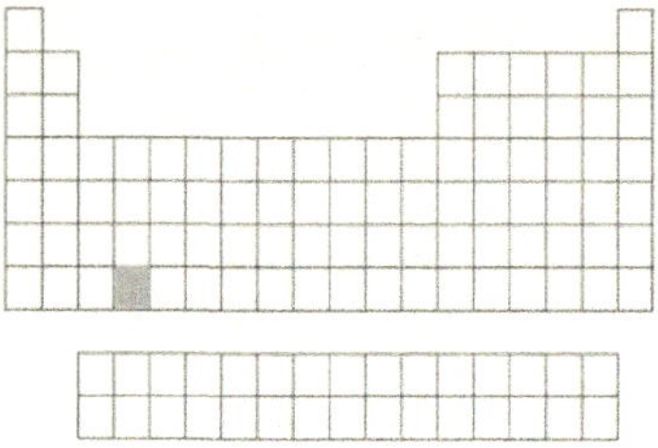

Its first, highly questionable sighting was in Dubna in 1964, where plutonium was bombarded with neon ions and the results analyzed using thermochromatography. These results may have been accurate, but they were altered over the years and lost much of their credibility. In 1969, Berkeley physicists bombarded californium with carbon and conclusively found element 104, results which were independently confirmed in 1973. A controversy arose, the competing claims being weighed by the International Union of Pure and Applied Chemistry (IUPAC), who, after careful consideration, gave the labs joint credit. This did not end the controversy. They then gave the transfermium elements temporary names which were simply Greek numbers, and spent a few decades trying to sort out the rival claims, and settle on permanent names that would not make anyone angry.

At long last, in 1997, rutherfordium got its name.

Named for the father of nuclear physics, who discovered the atomic nucleus (among many other things), rutherfordium was a name first proposed by the Russians for element 103, then suggested by Berkeley for 104, then assigned by IUPAC to 106, before coming back home to be 104 forever.

It has no uses.

Investigation

It was quite the most incredible event that has ever happened to me in my life. It was almost as incredible as if you fired a 15-inch shell at a piece of tissue paper and it came back and hit you…It was then that I had the idea of an atom with a minute massive centre carrying a charge.

-Ernest Rutherford

Rutherfordium is synthetic, radioactive, and technically a transition metal; but it's too large to exist outside of the few times it has been built. So do let's talk about Rutherford.

Ernest Rutherford, born in New Zealand in 1871, was an absolute powerhouse of a scientist. His contemporaries—even the smartest ones—admired him, and his credentials are legendary. Before his work, science thought the atom was indivisible. He discovered the nucleus; discovered and named the proton; found and identified alpha and beta radiation; named gamma radiation. He brought a fresh perspective to a debate on the age of the Earth by suggesting, correctly, that the Sun was radioactive, and on and on.

Unusually for the geniuses honored in this part of the table, Rutherford was also, by all accounts, a really nice guy. He made friends easily, and inspired and mentored a "who's who" in the physics world. Everyone who worked alongside him, and everyone who came after him, listed him as an inspiration for their greatest works. But with all of this larger-than-life legacy, still the most insightful monument to him is a bronze sculpture in Brightwater, New Zealand—near his birthplace—of him as a child, holding a mathematics book, getting ready to enthusiastically investigate the universe.

Look more closely at your world, is the message of rutherfordium. Explore the tiny things that are usually hidden. Unpack the stuff in your head that you thought was solid. Unfold the layers of assumptions and superstitions to see the heart of your thoughts. Look to the small things in your physical world, too; really look at a daisy, or a teacup, or a pebble. Be fascinated, and fearless, and curious. Investigate.

Dubnium

Element Number:	105
Weight:	262u
Normal State in Nature:	synthetic solid
Melting Point:	unknown
Boiling Point:	unknown

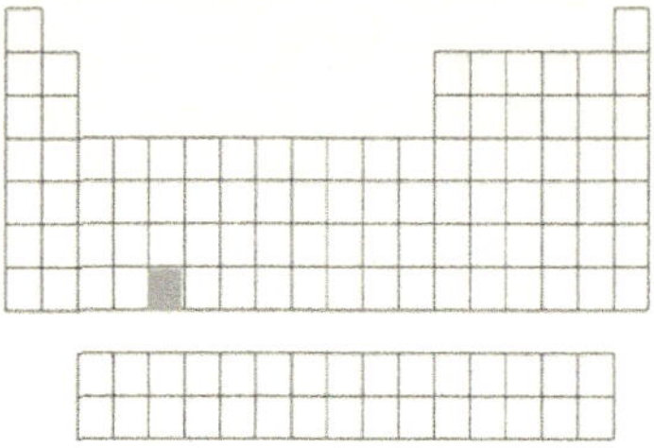

The "Transfermium Wars" was the rather dramatic name for the controversy between the Joint Institute for Nuclear Research, or JINR, in Dubna, USSR; and the physics lab at the University of California at Berkeley. They were racing to fill in the last several blank spaces in the periodic table, and the race was not always amiable. Element 105, for example, was observed in 1968 by the JINR team when they bombarded americium with neon ions. But then in 1970 the Berkelians threw nitrogen ions at californium, and achieved a similar result. The Dubna lab had delayed their publication in order to confirm their findings, Berkeley insisted that the discovery was theirs, and the IUPAC had their hands full. It was an absolute mess and it went on for decades.

Dubnium was finally settled on in 1997, after various concessions and rule-changes and a certain amount of childish behavior. Berkelians have been known to occasionally call it *hahnium* regardless, to protest the honor being given to their rival lab. Notwithstanding this, the Transfermium Wars are over, thank goodness.

Dubnium has no uses.

Identity

When you gonna make up your mind? When you gonna love you as much as I do?
-Tori Amos, *Winter*

Dubnium is synthetic and radioactive. Its atoms are rarely built, and they decay within 28 hours. So what is there to talk about?

Well, there are these two nuclear research laboratories. They have always had a healthy rivalry, although since they're in the USA and Russia…their rivalry has, on occasion, been a tad less than healthy. In 1969 the lab in Dubna discovered element 105, but hesitated to name it. Then the Berkeley lab discovered it in 1970. Dubna said, *That's ours, back off. We've decided to call it nielsbohrium.* But Berkeley came back with *You snooze, you lose, and we're calling it hahnium.* Neither would give an inch. A neutral international committee was formed, but their only immediate contribution was to suggest a third, uncontroversial name, *unnilpentium* (which just means 105-ium). This failed to calm things. The poor element was stuck with three names for years.

A German lab joined in. The controversy grew. It encompassed every element number over 100, and thus came to be called The Transfermium Wars. Finally, in 1997, the Berkeley lab grudgingly let the name of 105 get clean-slated and called dubnium to honor the city of their rivals. It was a hard sell. Rumor has it that there are still Berkelians who refuse to call it that. But so it goes.

Dubnium, therefore, is about identity. Who are we? Who decides? Do decades have to pass for us to find ourselves? Who do *we* think we are?

When you have drawn dubnium, you are being asked to define yourself. If this is a thing you struggle with constantly, wrestling to overcome your programming; and never sure, from day to day, whether you are strong or weak, happy or angry, male or female, important or invisible; be comforted with the knowledge that dubnium knows how you feel. It says: define what you are today. Tomorrow you may be different, and that's okay. And it's also okay to ignore those who don't call you by your chosen name.

Seaborgium

Element Number:	106
Weight:	269.128u
Normal State in Nature:	synthetic solid
Melting Point:	unknown
Boiling Point:	unknown

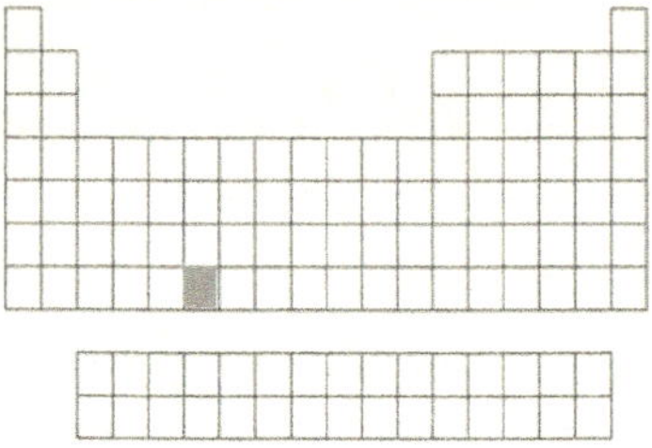

Element 106 was probably synthesized in 1971 by the Berkeley lab, using oxygen projectiles and a californium target. Their particle accelerator was shut down before they could confirm, to undergo extensive upgrades. When they finally attempted the experiment again in 1974, the findings were identical, but by then the lab in Dubna had only just published their discovery of 106 using lead bombarded with chromium.

As with every element in this part of the table, the rival claims exploded into a controversy between stubborn scientists, which was not resolved until 1997.

The IUPAC made a rule in 1994 that living persons could not have elemental namesakes. But they had also decided that Berkeley had the greater claim to discovery of 106, and the Berkelians were, before that rule was written, already resolved on naming this element after their beloved mentor, Seaborg, who did happen to still be alive, and working, at 85 years old. The IUPAC rule seemed to many to be bad form, and rather petty. It was eventually rescinded, and seaborgium is the result.

It has no uses.

Celebrity

you know I'm connected to the powers that be

-Reb Fountain, *Don't You Know Who I Am?*

One cannot study chemistry, atomic theory or nuclear history for five minutes without coming across the name of Glenn Seaborg. He was personally involved in the discovery of ten elements (including this one); he found more than a hundred transuranic isotopes; he figured out what actinides are, and where they belong in the table. He won a Nobel prize and over 50 honorary doctorates. He had the ear of every president from Truman to Clinton, and was the first person to have an element named after him while he was still alive.

He worked on the Manhattan Project, but was thoroughly against dropping the bombs on cities. After the war, he was an avid proponent of putting atomic technology to peaceful purposes. He was also enthusiastic about learning, and was one of the people involved in the 1970s attempt to revive our flawed education system. His list of credits can lead one to wonder if he ever slept, or if maybe he was secretly twins. Definitely someone worth celebrating.

If one wishes to know more about him, one is very much in luck: Seaborg kept a daily journal for decades, was co-writer of over 500 scientific papers, and had an open-book sort of lifestyle. He and his wife built their own house and were together for 56 years. He was important, but more importantly, he was *good*. He did not seek celebrity status, but he had it, nonetheless. He had—and has—a large and enduring fan base.

Celebrate your accomplishments. Sign your name big and flourishing on whatever your work may be. Take compliments gracefully, and pay them forward. Step into each day like there's a red carpet laid out before you.

You really are extraordinary and worthy of praise. Maintain your moral compass, but revel in your talents. Glenn stayed grounded his whole, glorious life; you can too.

Bohrium

Element Number:	107
Weight:	264u
Normal State in Nature:	synthetic solid
Melting Point:	unknown
Boiling Point:	unknown

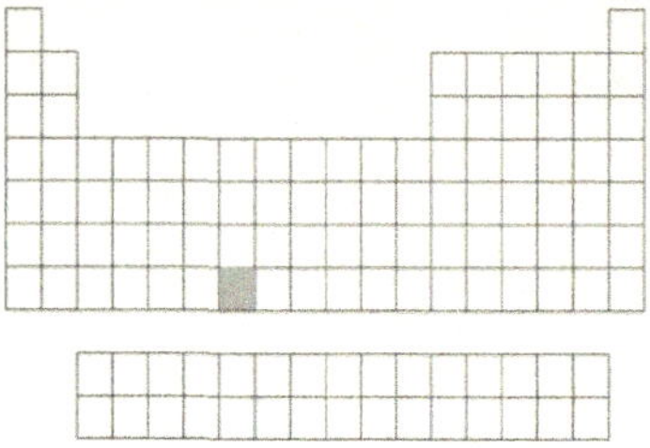

The German lab in Darmstadt was the first to definitively synthesize element 107 in 1981, by bombarding bismuth with accelerated nuclei of chromium. They produced five atoms. Their claim was much more convincing than that made by the Dubna team in 1976, who had bombarded bismuth with chromium as well as lead with manganese, producing what they believed to be 107 but could not substantiate. The German lab, once given the naming priority, very diplomatically suggested a name that the Soviets had previously wanted for 105, nielsbohrium.

The only objection that the IUPAC had was in including the man's first name; and so, despite complaints that it would be confusing in its similarity to "boron," in 1997 bohrium became official.

Beyond pure experimentation, bohrium has no uses.

Wholeness

If you can see a thing whole, it seems that it is always beautiful. Planets, lives…
-Ursula K. LeGuin

Since bohrium is on the "only exists officially" list, let us discuss Bohr himself.

Of the many—oh so many—accomplishments adorning the life of Niels Bohr, the one we will focus on is his concept of complementarity: Every item in quantum mechanics has two properties or forms which cannot be observed at the same time. Their determination depends on the experimental framework. Thus one can measure the position of an electron, or its momentum, but not both; or see how light behaves as a wave, or as a particle, but not both. The information cannot be reincorporated after the fact, either; one must be able to embrace both pictures as true without trying to squash them into one picture.

Yes, he also invented the Bohr model of an atom, which has been supplanted in accuracy but not in usefulness. And yes, he convinced the Danish government to let him establish an Institute of Theoretical Physics, which became a favorite haunt of every early-20th-century physicist we've ever heard of; and yes, he personally inspired great minds from Rutherford to Heisenberg; the latter of whom described him as "primarily a philosopher, not a physicist." This leads us back to complementarity, which has an almost Taoist flavor. Let Things Be What They Are, it seems to say. When there are dozens of conflicting facts that can all be proven, do not try to bend them to suit each other. The Whole can be known by a mind that is open.

Bohrium is a message of making peace with things as they are. The world is full of contradictions, and so are we; you yourself can be both pleased and frustrated; both happy and sad. The planet is simultaneously screwed and blessed; humanity is kind and it is vile. You don't need to cut one short to see the other. Observing without censoring is the path to wholeness.

Hassium

Element Number:	108
Weight:	277u
Normal State in Nature:	synthetic
Melting Point:	unknown
Boiling Point:	unknown

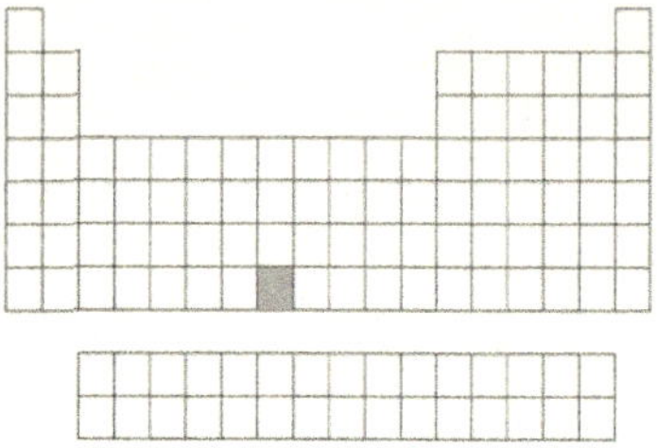

The synthesis of element 108 was attempted by Oganessian and his team at JINR no less than three separate times: In 1978, 1983 and 1984. None of these were unambiguously successful, and Oganessian included that information in his reports. But, also in 1984, the Gesellschaft für Schwerionenforschung (or GSI) in Darmstadt, Germany bombarded lead with iron nuclei, which conclusively resulted in element 108. The IUPAC examined both labs' findings, and decided that priority of discovery belonged to Darmstadt. The Russians graciously agreed.

Darmstadt is in the state of Hesse, which is called Hassia in Latin. Hassium was not officially named until 1997, as it was caught up in the Transfermium Wars, but when the dust finally settled, hassium's name had survived the turmoil.

Only a few of hassium's properties have been observed, as it exists so rarely and briefly. Most of what is known about it is conjecture.

It has no uses.

Loyalty

Following my muse has worked out pretty well so far. I can't see any reason to change the formula now.

-Chris Van Allsburg

During the Transfermium Wars, which took place roughly from the 1960s to the 1990s (and are really just a glorified way to describe physics labs quibbling over the names of superheavy elements), the Gesellschaft für Schwerionenforschung (or GSI) in Darmstadt, Germany, was not behindhand in its own claims. Filling in this area of the periodic table was a race during those decades, and element 108 was one of many times that two labs published their discovery of it simultaneously. The international committee who was in charge of keeping things civil decided that the German lab had the most clear-cut results, and the lab joyfully slapped the name hassium onto the element, after their beloved state.

Darmstadt is in the state of Hesse. Back before there was a united Germany, there were sixteen small countries that considered themselves independent of all the others; despite sharing a similar heritage, language, cuisine and aesthetic, they were nonetheless distinct. And though the states had all been unified in 1871, the sense of place has persisted. Hesse still has its local customs and laws. There is a Hessian anthem and a Hessian crest. To be loyal to one's origins, one need not be toxic about it. One might simply discover an element, and name it after one's homeland. The element may only exist when synthesized, and last, at most, ten seconds; but the name is forever.

Loyalty is the byword of hassium; loyalty to your place, your people and your foundation. Think about your origins. Not just the origins of your physical being, but of your convictions and your ideals. Where did those come from? Who planted those seeds? Where were your best memories formed? Where did you decide who you wanted to become? Take a moment to appreciate that place and those people. If you cannot immortalize them in an internationally recognized elemental nomenclature, at least send them good energy. Remember your roots.

Meitnerium

Element Number:	109
Weight:	278u
Normal State in Nature:	synthetic
Melting Point:	unknown
Boiling Point:	unknown

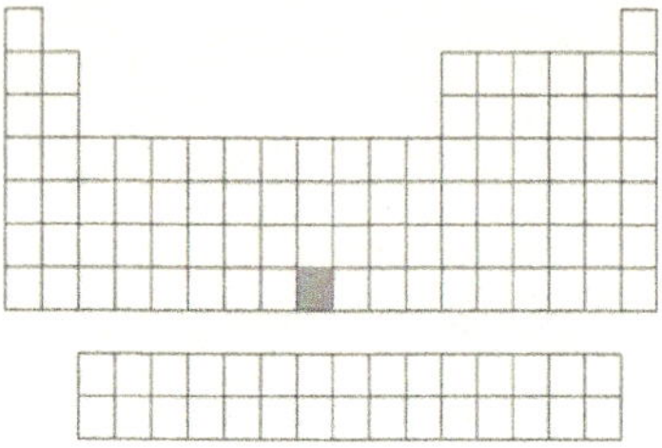

The GSI in Germany bombarded bismuth with iron in 1982 and created a single atom of element 109. The JINR repeated the experiment in 1985, and confirmed their findings.

Of all the elements whose names were under dispute during the Transfermium Wars, meitnerium was the only one that no one argued. Proposed by the team at GSI, meitnerium would be to honor Lise Meitner, a powerhouse in the physics world and one of the discoverers of nuclear fission; arguably the most important revelation in human history.

Meitnerium has a very short lifespan. Most of its isotopes last for a handful of microseconds. Therefore, aside from its atomic number, it has no measurable traits at all. It also has no uses.

Recognition

She had had one idea firm in her head and had held to it when every one thought she was wrong; and now, when everybody knew she had been right, just for a minute or two she didn't want to do any talking.

-Arthur Ransome, *Swallows and Amazons*

Meitnerium exists in that weird, synthetic, short lived area of the table where a discussion of chemistry is impossible. But Meitner herself is worth a look. Her achievements are legion. Alongside her long-time lab partner, Otto Hahn, Meitner discovered nuclear fission in 1938—arguably the most significant discovery of all time. In 1944, the Nobel Committee acknowledged this by giving the prize…to Hahn alone.

This lack of recognition was already a pattern. Lise Meitner is the poster child for the exploitation of female scientists. She frequently went unpaid, underpaid, passed over and dismissed. It must have become something of a joke that a Nobel prize was withheld from her. She was nominated 48 times, in both physics and chemistry, but was always disregarded. Eventually she was simply invited to exclusive Nobel laureate functions because the alumni were tired of waiting.

They weren't the only ones with a vast respect for this woman. Einstein praised her; Planck and Seaborg loved her to pieces. She has namesakes in craters, on both the Moon and Venus. And finally, in 1997, this element was named in her honor. So while recognition may be a long time coming, and lack of it can feel downright insulting…Meitnerium is a reminder that it cannot be withheld forever.

Today is a day to acknowledge and celebrate your accomplishments. You have done some amazing things, and you deserve to feel proud. Self-promote. Add fresh successes to your resume. Treat yourself to dinner. Also, look around you, at others in your life whose achievements have gone unpraised. Show them that you see them. Reward them for their hard work, if only with words. It is surprising how much of a difference that can make.

Darmstadtium

Element Number:	110
Weight:	281u
Normal State in Nature:	synthetic
Melting Point:	unknown
Boiling Point:	unknown

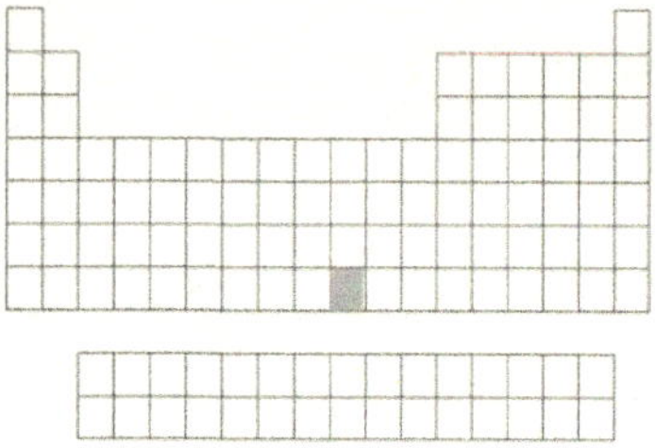

Hofmann and his team, Armbruster and Münzenberg, at the GSI in Darmstadt, synthesized element 110 in 1994. They bombarded lead with nickel nuclei and detected a single atom. Within that series of experiments, they identified nine more 110 atoms.

The labs in Dubna and Berkeley had tried to synthesize 110, but their results were not conclusively successful. This did not prevent them both from putting forward name suggestions, all of which were ignored by the IUPAC. The Darmstadtians found it, so they got naming priority. They went with darmstadtium.

Darmstadtium has no measurable traits. It is known to be radioactive, and that is all. It has no uses.

Experimentation

"In him I see already the fire, dust and air."
"Well then we'll just add water and stir."

-Auntie Mame

Darmstadtium holds together for no more than a few seconds, so its properties can only be guessed at. Let us discuss its name instead.

Darmstadt is officially *the* preeminent German *Wissenschaftsstadt*; as you may already know, this means "City of Science." And no wonder. It is home to several universities, institutes, space centers, and, not least, the GSI: The organization which is responsible for discovering—or rather, synthesizing—several superheavy elements, including darmstadtium itself. They have a particle accelerator. It's really shiny.

Some towns get honored in the periodic table because they happened to be near the mine where a given element is found. Ytterby, Sweden and Strontian, Scotland are examples of this. But some towns earn their immortality within chemistry nomenclature by being centers of learning, of discovery and experimentation; and Darmstadt is certainly one of these. Whether looking into the stars, into the future or into the hearts of atoms, Darmstadt is here for it. Trying new things. Experimenting.

Darmstadtium means it is time to try new things for the purpose of creating change. Experiment with your life. Eat strange foods. Change your workout routine—or, if you're like me, try even having a workout routine. Get your hands dirty and avoid pointless theorizing. Look outward, and inward, and find things to tinker with. Your results may surprise you.

Roentgenium

Element Number:	111
Weight:	272u
Normal State in Nature:	synthetic
Melting Point:	unknown
Boiling Point:	unknown

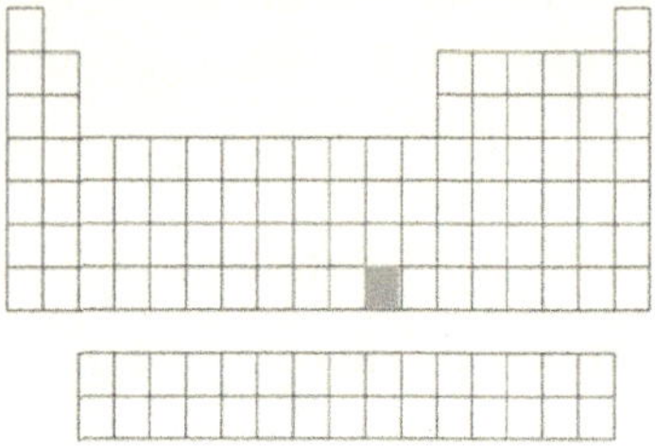

This time, Hofmann and his team at the GSI bombarded bismuth with nickel and detected three atoms of element 111 before it dissolved. This was in 1994, long after the failed 1986 attempt made by the JINR.

Again, in 2002, the GSI repeated the experiment as well as the success, and sealed their right to suggest a name.

In 2004 they finally settled on roentgenium, to honor Wilhelm Roentgen, the discoverer of x-rays.

Roentgenium has no measurable traits and no uses. It might behave like the other noble metals in its series, but it does not last long enough to find out.

All we know for sure is that is does not emit x-rays.

Irony

Things really started to improve in 1895 when German physicist Wilhelm Roentgen …discovered x-rays. This made it possible for the first time for doctors to look inside living human beings and spot problems such as cancerous tumors, which, as it turned out, were caused by exposure to x-rays. So this invention came along just in the nick of time.

- Dave Barry, *I'll Mature When I'm Dead*

Roentgenium is yet another element that only exists as a few short-lived atoms by dint of much human effort. Let us instead learn about Roentgen.

Wilhelm Roentgen's life is a study in irony, of the colloquial kind. He was expelled from high school for a prank he didn't pull; as a result, he had to hunt for a university that would take him without a diploma. Four years later, he had a PhD in mechanical engineering. He planned to emigrate to the US; he had a job waiting and even had his boat tickets…when WWI broke out. He never did make that move. He never patented any of his amazing discoveries because he felt they belonged to mankind; this left him consistently broke. "X-rays" was meant to be a temporary name, by the way; a placeholder. But, like so many things, it stuck. Incidentally, one of the few things known about roentgenium is that it does *not* emit x-rays.

Roentgen is considered the father of modern radiography, and though his work may have given him the cancer that eventually killed him, it has also saved countless lives—and improved the quality of even more.

Isn't it ironic?

Sometimes you have to laugh, if only to keep from crying. Life is handing you some weird, contradictory baggage to work on. Thank goodness you have a sense of humor, right? Just remember that sometimes life gives you lemons, and then water, and then sugar, and a juicer…but then you cannot find a glass for love or money. Whatever you're going through, take comfort in the fact that you'll have a story to tell. We've all been there. We'll sing along.

Copernicium

Element Number:	112
Weight:	285u
Normal State in Nature:	synthetic
Melting Point:	unknown
Boiling Point:	unknown

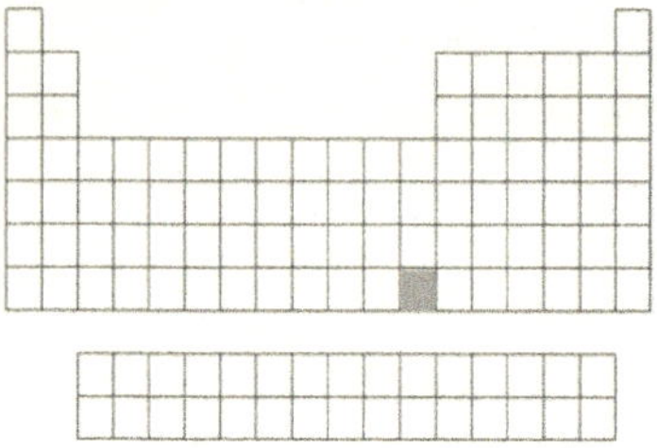

In 1996, the Gesellschaft für Schwerionenforschung was on a roll. This time, Hofmann and his team fired zinc nuclei at a lead target and produced a single atom of element 112. They repeated this feat in 2000, and had their results confirmed by the lab in Japan, Riken, in 2004 and 2013.

At first the IUPAC felt the data was insufficient to confirm discovery, but once more was known about the behavior of neighboring isomers, the conclusion became clear. The committee then asked for a name suggestion, and they offered copernicium, to honor a scientist "who changed our view of the world."

The name was ratified on February 19th, 2010; Copernicus' 537th birthday.

Copernicium's properties are a mystery. It has no measurable traits and no uses. All is conjecture.

112

Perspective

While I am writing reports I am freethinker und atheist, but out here in the Rukh I am more than Christian. I am Pagan also.

Rudyard Kipling, *All the Mowgli Stories*

Only a handful of copernicium atoms have ever existed. It has no measurable traits, but it has a name…

Copernicus arrived early for the Renaissance. At the turn of the 15th century, he was a mathematician, linguist, economist, physician and astronomer. Oh yes, astronomer. In an era when ancient knowledge was all but forgotten, Copernicus was the first to rediscover some home truths; the most famous being, *we are not the center of the universe*. This was a revolutionary idea, especially for someone who was a devout Catholic. He observed the facts and wrote his paper. But he delayed printing it, even though he was urged to, fearing to anger the church. Legend has it that he finally, aged 70, gave in, sent his manuscript to the printer…and died the day it was delivered, with the book in his hands. At first the blowback was fairly weak. It wasn't until sixty years later, when Galileo expanded on the theory, that the church went nuts. Whatever religion says, the Copernican Model is the correct one. Except that he didn't take it far enough. Not only is the Earth not at the center of the universe, but neither is the Sun. Neither is the black hole that holds our galaxy together. Neither is the supercluster of galaxies that we belong to. In actual fact, there probably isn't a "center of the universe" at all, but if there is, it is not us.

Copernicium is here to remind you that it is far better to change your mind than to close it. Old ideas, especially regarding your own importance, are comfortable and safe; but they often require you to dismiss or avoid so many experiences, so much knowledge and awe and joy. To cling to your ego rather than explore and discover is…well, you might as well be buried alive. So shed any notion you have about being the center of the world, and instead grasp the much more liberating idea that you don't have to be. Open your eyes, and learn things.

Nihonium

Element Number:	113
Weight:	286u
Normal State in Nature:	synthetic
Melting Point:	unknown
Boiling Point:	unknown

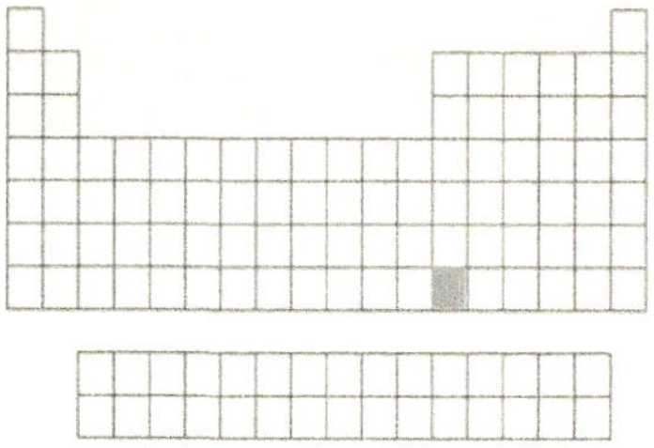

The JINR and the Berkeley lab set element 113 aside temporarily, because the odd-numbered heavy elements are more difficult. While they were wrestling with 114, the Riken Nishina lab in Wakō, Japan, quietly set to work. Morita and his team repeated experiments to confirm elements 108 and 110-112, then used the same methods to search for 113. They bombarded bismuth with zinc for several months.

They were successful.

Meanwhile, the JINR-LLNL (Dubna and Berkeley) collaboration claimed that they had achieved 113 as part of decay chains for the even heavier elements they'd been working on. Their results were weaker than those of the Riken lab, so Morita's team got to put forward a name.

By naming this new element nihonium, Morita's people were doing many things. They were thanking the people of Japan, as the lab was publicly funded; nurturing a sense of national pride and achievement; and giving honor to Ogawa who, back in 1908, had had his claim of the discovery of *nipponium* rejected, that element being named *rhenium* instead. Nihonium was the realization of a hundred-year-old dream.

Being a superheavy element, it has no uses or observable traits.

Incognito

Would you like my mask? Would you like my mirror?
Cries the man in the shadowing hood

-Loreena McKennitt, *The Mask and the Mirror*

It only exists for a moment. It is the last of the boron group, but has no traits that can be measured or observed. It is unknown and unknowable. It is shrouded in synthetic, radioactive obscurity.

Its name, on the other hand, is at least an open fact.

Nihonium is the only element whose discovery is credited to an East Asian team. That lab decided, therefore, to honor their country in the naming, and Nihon is one of Japan's real names. Hence, Japan itself is the thing to talk about when nihonium is brought up—particularly its relation to the rest of the world. To those in the West, Japan has always been mysterious. There have been swathes of time when the small but sophisticated island nation simply did not allow foreigners to approach; they neither needed nor wanted anything from the outside world. This made Japan seem almost unreal, like an untouchable faerie land. And when westerners finally did get to interact with the culture of Nihon, they found it still profound, multilayered, and somehow hidden from them. The history is deep, the customs are complex. Outsiders can try, but they will never really see the whole truth of the place. It will always be terra incognito.

With nihonium in your spread, it is best not to try too hard to wrap your mind around the thing you are confronting. Whether a person or a concept, a place or some inner part of your self; if it seems to be eluding you, let it. If it shies away from questioning, stop asking. Sometimes a mystery needs to stay a mystery, and your respect for that is noble and just. You can embrace whatever-it-is, enigma though it may be. You do not need to understand a thing before you can love it.

Flerovium

Element Number:	114
Weight:	289u
Normal State in Nature:	synthetic
Melting Point:	unknown
Boiling Point:	unknown

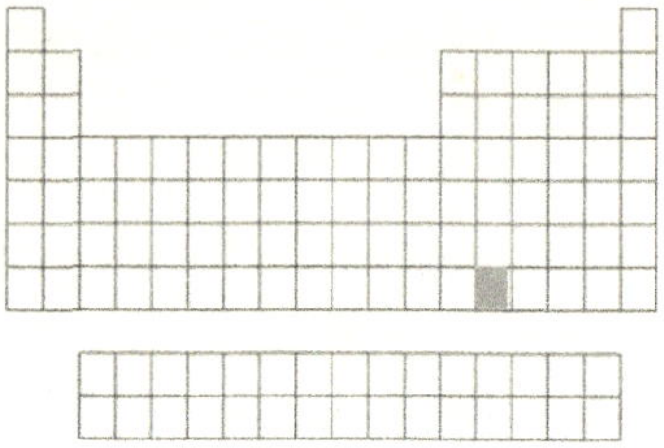

Elements 107 to 112 were synthesized using cold fusion: The tendency of atoms, when they come together hard enough, and have a stable enough initial configuration, to merge. This appeared to stop working at element 113, though, and it seemed that a different method would be needed. Oganessian's team at JINR attempted *hot* fusion, which requires an actinide target. They bombarded plutonium with calcium in 1998 and detected a single atom of 114. A year later, they repeated it with a different isotope of plutonium, and got *two* 114 atoms. Berkeley confirmed the finding in 2009, and Darmstadt in 2011. Dubna's suggested name for it, flerovium, was meant to honor the Flerov Laboratory of Nuclear Reactions, which in turn was named for Georgy Flyorov, the Soviet physicist.

It's unclear whether flerovium is a metal or a gas. So far, about 90 atoms have been created. The most stable one has a half-life of less than two seconds. Flerovium has no observable properties, only predictions. It has no uses.

114

Susurrus

Tiffany liked the *taste* of the word. It made her think of mysterious people in long cloaks whispering important secrets behind a door: *susurrusssusurrusssss…*

-Terry Pratchett, *The Wee Free Men*

Flerovium is interesting in that, unlike its neighbors, it *has* been observed chemically…but the results were so fleeting and elusive, scientists cannot agree whether it behaved more like a metal, or a noble gas. For such a point of confusion to exist must create a lot of static in physics circles.

In keeping with the theme of obscurity, flerovium is not named for Georgy Flyorov, as one would expect. He did, after all, discover spontaneous fission in 1940; and it was through his agency that the Soviet Union started its own nuclear bomb program. He had noticed a suspicious silence on this topic from his fellow physicists in allied countries, and deduced—correctly—that the silence meant, not absence, but secrecy. Whispers behind closed doors…

But no, this element is in fact named for the *lab* that was named for the man. Yet another way to keep things vague and blurry. A further susurration regards the discussion of the fabled island of stability that is conjectured to exist in this part of the table. The theory goes: The radioactive element half-lives become shorter and shorter the higher we go, but somewhere in here there might just be isotopes that, if we succeeded in synthesizing them with the proper number of neutrons, would last longer. The math supports it, but it is still only an idea, to be talked about in hushed, serious voices…*susurrussusurrus…*

Flerovium is about listening to whispers. The information you need is not going to come at you loud and obvious; but in strange, sidling, secretive ways. Pay attention. Listen at keyholes. Loosen your grip on sharply defined communications and instead let your mind absorb the buzzing in the background, the undercurrent of things unspoken but hinted at. This is where you will find what you seek.

Moscovium

Element Number:	115
Weight:	288u
Normal State in Nature:	synthetic
Melting Point:	unknown
Boiling Point:	unknown

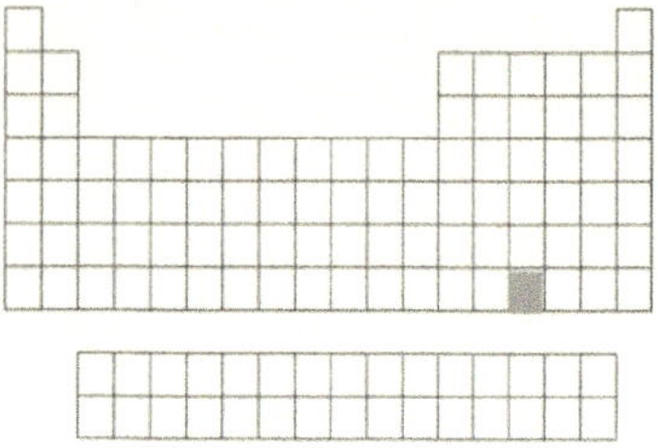

In 2003, the Russians and Americans laid aside old differences and joined forces to achieve the synthesis of the last few elements that would make the table whole. Their first challenge was element 115. They bombarded americium with calcium and achieved four atoms of the new element, which lasted for nearly 100 milliseconds. The IUPAC was hesitant to concede the discovery, though, until it could be repeated. Which it was, in 2013, by the team in Darmstadt. The physics lab at Lund University in Sweden was doubtful of these results, and those of tennessine, which were connected. They may still be doubtful. But the official naming happened, regardless.

Moscovium was named in 2017, for the city of the lab in which it was synthesized.

Moscovium has no measurable traits or uses.

115

Silence

This page intentionally left blank.

-every standardized test

This
is not
a time
for words.

Livermorium

Element Number:	116
Weight:	292u
Normal State in Nature:	synthetic
Melting Point:	unknown
Boiling Point:	unknown

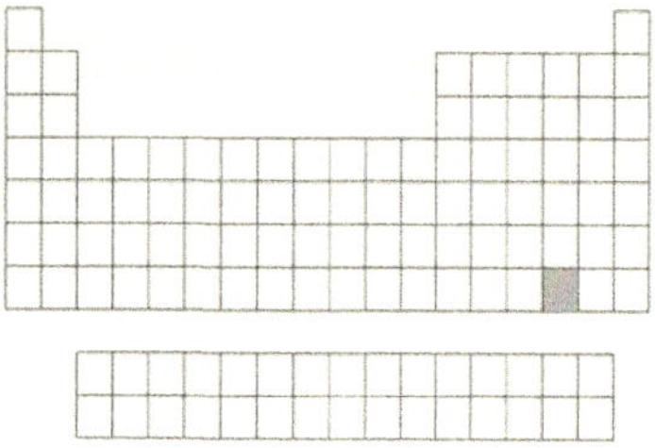

Hulet and his team at Berkeley attempted synthesis of 116 way back in 1977, but it was unsuccessful. Oganessian tried too, in 1978; Armbruster took a swing at it in 1985. In the 1990s, Berkeley and Darmstadt began a collaboration using stronger calcium beams, actinide targets and *much* more sensitive equipment. This turned out to be a really great idea. But even they were not the discoverers. That honor goes to Dubna, in 2000, where they bombarded curium with calcium and detected a single atom of 116, confirmed later with repeats of the experiment and through careful observation of decay chains. Much of their work was in collaboration with the Berkeley lab, so when the IUPAC approved the results, they graciously offered livermorium as a name, to make it a pair with moscovium.

Livermorium is named for the Lawrence Livermore National Laboratory, which was in turn named for both Ernest Lawrence and the city of Livermore, California; which was named after a naturalized Mexican rancher of English birth, who had lived in the first half of the 19th century.

It has no observable traits or uses.

116

Absurdity

What is, is, and what might have been could never have existed.

- Edward Gorey, *Ascending Peculiarity*

Livermorium only exists, briefly, when great effort is put into making it. Its name is worth talking about, though, because of the ridiculousness of it all.

It was named for the Lawrence Livermore National Laboratory in California, which was founded in the 1950s to focus on nuclear weapons research. Don't let the name mislead you; there never was such a person as Lawrence Livermore. There was an Ernest Lawrence (who has his own elemental namesake, 103), and there is a city called Livermore. The lab is named for them both.

Livermore, the city, was named after Robert Livermore. His was an interesting life, in which he went from being a stonemason's apprentice in Essex to a soldier in the Peruvian navy to a naturalized Catholic Mexican ranch owner and father of eight. It's all just a bit absurd, frankly. Also, he never went anywhere near the area that became the town that bears his name; it was simply founded by someone who admired him.

Therefore, it can truly be said that the etymology of this element can be traced from a lab, to a city, to a guy who liked a rancher, to said rancher; who, through all his rich and convoluted life experiences, may or may not have even heard of physics. Or chemistry. Or the periodic table. With all the hundreds of unsung geniuses who have worked in the field and could have been given this extremely limited honor…oh man.

Livermorium is a sign that you just gotta laugh. Whatever is going on in your world, it is absolutely ridiculous. You can decide whether it's "If I don't laugh I'll cry" ridiculous or "You can't make this shit up" ridiculous, but whatever flavor of stupid this is, just run with it. Embrace the silliness. Post it in a thread. Tell your friends. Laugh till your stomach hurts. Do not try to change it, and don't bother trying to make it make sense. Sometimes, life is just plain absurd.

Tennessine

Element Number:	117
Weight:	294.211u
Normal State in Nature:	synthetic
Melting Point:	unknown
Boiling Point:	unknown

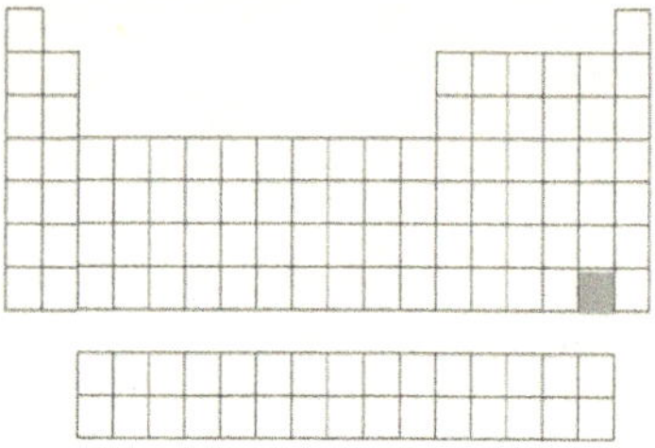

By 2004, every element in the periodic table had been created and identified, save one. Element 117 was particularly difficult. The JINR proposed a collaboration with the Oak Ridge National Laboratory in Tennessee, because they theorized that what was needed for the experiment was berkelium, and berkelium was made at Oak Ridge and nowhere else.

Berkelium is a byproduct of californium production, so the teams waited for a commercial order to come in and kickstart the process. In 2008 this finally happened, and the collaborating labs, who by now included Vanderbilt University and Berkeley, leaped into action. Armed with 22 milligrams of berkelium worth $600,000, the clock was now ticking. With a half-life of 330 days, there was no time to lose. Of course it had to be cooled for 90 days, then purified for 90 days.

And finally, it had to cross the Atlantic Ocean five times in four days, because Russian customs officials are sticklers for paperwork.

In the end, the experiment was a success, and the element named for Tennessee, where Oak Ridge made it all possible.

Tennessine has no observable traits or uses.

117

Rumor

Man's greatest flaw: The illusion of control.

-Reginald Hargreeves, *The Umbrella Academy*

Tennessine was the last element to be officially synthesized, and even then, there were brilliant people peering at pages of data for ages before and after its confirmation, arguing. When something exists for only a blip, as a handful of atoms, one cannot simply point and say, "There it is." One must follow the decay chains, looking for the shadow or the footprint of the elemental cryptid. And before that, one must have the materials to even make the attempt. Several laboratories joined forces to this end. Element 117 was the theoretical product of bombarding berkelium with a calcium beam, but how to get the berkelium? There was, at the time, only one place to go: Oak Ridge National Laboratory, in Tennessee. They could create berkelium as a biproduct of californium (for which there is niche commercial value), but it was incredibly expensive and the lab had no orders to fill. So the team waited…Listening for a rumor that production had begun again. When it was, the berkelium was purchased and sent in lead cases to Dubna in a hurry, because when your materials are radioactive you have a limited time to use them. They were improperly labeled, though, and sent back. Twice. Our physicists were frantic--according to rumor. But the story has a happy ending, and Oak Ridge's home state was honored in the naming. Still, tennessine remains shadowy and vague; a set of numbers on a spreadsheet rather than a real halogen. A story. A rumor.'

Tennessine wants you to channel your inner cryptid. Haunt forests, either literally or otherwise. Be wild and free in the mist of obscurity, and see what happens. Leave footprints and see who follows. Perhaps nobody will, and that's okay; you exist independently from an audience. So, do your thing. Flit between the trees. Leave an unexplained trail of seashells or flower petals. Let those who find you go back to civilization babbling about the experience. Be a rumor.

Oganesson

Element Number:	118
Weight:	294u
Normal State in Nature:	synthetic
Melting Point:	unknown
Boiling Point:	unknown

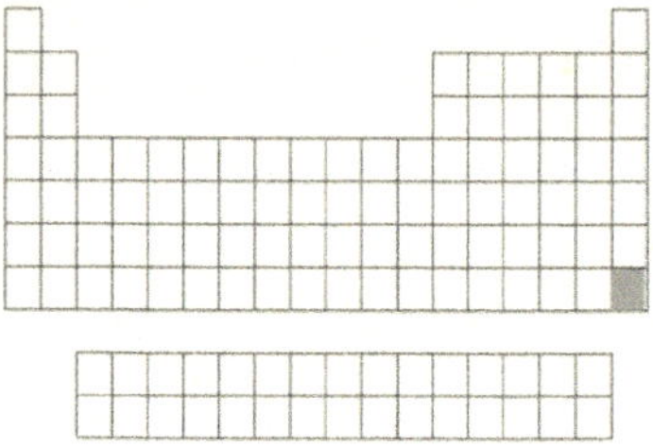

Oganesson has always been one of the most anticipated of the superheavy elements. Its position at the end of the table, and in the group of noble gases, has been exciting imagination since the table was conceived. Thomsen was speculating about it in 1895.

In 1999, Berkeley claimed to have synthesized it; but their results were unrepeatable, so they retracted their claim. In 2002 a confirmable report came from Dubna, where Oganessian and his team, including some Berkelians, witnessed a genuine sighting of element 118. They delayed their report until they could confirm by repetition, and in 2005, by bombarding californium with calcium, they were sure they had it.

The IUPAC was not so sure. In 2011 they stated that the results were inconclusive. But the labs kept at it, and by 2015 they had enough repeat success to declare 118 found.

Oganessian was asked to leave the call discussing the name of this final element, and everyone, all at once, suggested *oganesson.* That gave it the *-on* suffix appropriate to noble gases, and honored a man who had just spent 60 years leading the field in superheavy elements research. He is the second and last person to have an elemental namesake while still alive.

Oganesson has no uses and no observable traits.

At least for now, it is the end of the periodic table.

Death

How about not equating death with stopping?

- Alanis Morisette, *Thank U*

The final element in the periodic table as it stands today is of course a synthetic one. Theories abound as to its properties; it may be a noble gas, yet somehow reactive. It may be a solid. It might remain a mystery forever, because it is too heavy for this world.

Like death.

But the underlying truth of oganesson is that no one can say that it really is the heaviest element possible. There could be an eighth period waiting to be discovered, and who knows what those atoms will be capable of? What future scientists will be honored in their names? Yuri Oganessian is recognized here, the second and final person to have an elemental namesake while he is still alive; and as the head of the JINR in Dubna *and* the world's leading researcher in superheavy elements, he surely deserves it. Right now he can say that the ultimate element bears his name. Right now, we can think of death as the ultimate end of each living thing. But science is impressive and existence is strange, and the soul could very well be eternal, transcending all this crude matter. Like thoughts, or music. Or love. And when part of us dies, we do well to remember that we are more than any of the things we have ever been.

This is the end.

Except that there is no such thing. Death is transformation, which both physics and metaphysics agree is a universal experience, though they may differ on the details. They both, however, hold it to be a necessary process. Painful sometimes, beautiful sometimes, but always, always inevitable. You cannot stop it or fight it; but you can take its hand and step forward, into...whatever comes next. There are more adventures beyond the atoms you know. There is more to you than the part that you are leaving behind. Death foretells your next great and radical change, and we all look forward to it with wild anticipation. Be brave.

Made in the USA
Las Vegas, NV
12 July 2025

24832098R00146